THE NATIONAL INSTITUTES OF HEALTH
1991–2008

Also by the author:

Arrhythmias

Mergers of Teaching Hospitals in Boston, New York, and Northern California

Governance of Teaching Hospitals: Turmoil at Penn and Hopkins

Specialty Care in the Era of Managed Care: Cleveland Clinic versus University Hospitals of Cleveland

You and Your Arrhythmia: A Guide to Cardiac Arrhythmias for Patients and Families

Selling Teaching Hospitals and Practice Plans: George Washington and Georgetown Universities

THE NATIONAL INSTITUTES OF HEALTH
1991–2008

John A. Kastor
*University of Maryland School of Medicine
Baltimore, MD*

2010

UNIVERSITY PRESS

Oxford University Press, Inc., publishes works that further
Oxford University's objective of excellence
in research, scholarship, and education.

Oxford   New York
Auckland   Cape Town   Dar es Salaam   Hong Kong   Karachi
Kuala Lumpur   Madrid   Melbourne   Mexico City   Nairobi
New Delhi   Shanghai   Taipei   Toronto

With offices in
Argentina   Austria   Brazil   Chile   Czech Republic   France   Greece
Guatemala   Hungary   Italy   Japan   Poland   Portugal   Singapore
South Korea   Switzerland   Thailand   Turkey   Ukraine   Vietnam

Copyright © 2010 by Oxford University Press, Inc.

Published by Oxford University Press, Inc.
198 Madison Avenue, New York, New York 10016

www.oup.com

Oxford is a registered trademark of Oxford University Press

All rights reserved. No part of this publication may be reproduced,
stored in a retrieval system, or transmitted, in any form or by any means,
electronic, mechanical, photocopying, recording, or otherwise,
without the prior permission of Oxford University Press.

Library of Congress Cataloging-in-Publication Data
Kastor, John A.
The National Institutes of Health, 1991–2008 / by John A. Kastor.
p.   ;   cm.
Includes bibliographical references and index.
ISBN 978-0-19-973799-4 (alk. paper)
I. Title.
[DNLM: 1. National Institutes of Health (U.S.)—History.   2. Biomedical Research—history—United
States.   3. United States Government Agencies—history—United States.   4. History, 20th Century—
United States.   5. History, 21st Century—United States. W 20.5 K187n 2010]
RA11.D6K37 2010
610.72—dc22
2009032154

To Bob Lehrer

"Every day in this country a life is saved because of the breakthrough discoveries at NIH."—Donna E. Shalala, Ph.D., President, University of Miami; Secretary for Health and Human Services (1993–2001)[1]

The "most brilliant social invention of the twentieth century."—Lewis Thomas, M.D. (1913–1993). Dean, New York University School of Medicine; Dean, Yale School of Medicine; President, Memorial Sloan-Kettering Institute[2]

"The noblest activity of our government."—Eugene Braunwald, M.D., Hersey Distinguished Professor of Medicine, Harvard Medical School; NIH trainee and investigator (1955–1968)[3]

"The most successful public/private partnership in U.S. history."—John T. Potts, M.D., Jackson Distinguished Professor of Clinical Medicine, Harvard Medical School; NIH trainee and investigator (1959–1968)

"The best money the government spends is for medical research."—John E. Porter, member, House of Representatives (IL, 1980–2001); member, Subcommittee on Labor, Health and Human Services and Education of the Committee on Appropriations[4]

"Big place, lots of smart people."[5] Emblazoned on NIH coffee mugs.

CONTENTS

Preface	*xi*
Introduction	*xiv*
1. Intramural Research Program	*3*
2. Extramural Research Program	*31*
3. The Institutes	*41*
4. The Centers	*105*
5. Finances	*142*
6. Congress and Advocates	*154*
7. Directors	*161*
8. Controversies	*192*
9. Conclusions	*211*
Appendices	*215*
A. Interviewees	*215*
B. Acronyms	*233*
Notes	*236*
Index	*265*

PREFACE

THIS IS A book about the premier organization for performing and funding biomedical research in the United States. It is owned by the people of the United States and is an example of how an organization, although part of the federal bureaucracy, can function admirably for the benefit of all.

The book describes what happened at the NIH from 1991 to 2008, during the terms of directors Bernadine Healy, Harold Varmus, and Elias Zerhouni. I started the work by interviewing Zerhouni, then the NIH director, on January 24, 2007. Although the body of the text and most of the reporting ends in 2008, I have included a few of the most newsworthy events at the NIH since the end of 2008, such as the increase in the NIH budget early in 2009 and Barack Obama's lifting of the Bush administration's limits on stem cell research in March. By the time this book is published, other important changes will have occurred at the NIH and in the support it receives from the federal government. For the most part, I identify people by the jobs they were holding when the interviews were conducted.

The scientific accomplishments of the NIH during the 17 years covered by this book have been extraordinary. Of great importance also to the future of biomedical science has been the training of thousands of biomedical scientists that the NIH has supported.

In this book, however, there is relatively little medical science, except for a few examples where I attempt to emulate *Scientific American*. I am not a basic scientist and am unqualified to explain most of the research conducted at the NIH.[*] Rather, I write about how the NIH operates, its problems, its finances, its politics, and its

---

[*] The NIH has developed a comprehensive website with detailed information about what each of the institutes and centers do. Start at http://www.nih.gov/

structure within the federal government, in many instances through the people who work there. Readers who are knowledgeable about the NIH will observe that the book is not a comprehensive report on all that occurred at the NIH during these 17 years. Part of my motive here has been to keep the report reasonably concise.

Coverage of the 27 institutes and centers, the fundamental elements of the NIH, is not equal. I concentrated on those I thought would be of particular interest to readers and, not incidentally, to me. Generous space is devoted to HIV/AIDS in the National Institute of Allergy and Infectious Diseases; the problems conducting research in clinical cardiology in the National Heart, Lung, and Blood Institute;* how the NIH evaluates which grants to fund; and the unique features of the Clinical Center, the NIH's hospital.

Here are a few of the stylistic rules that I have followed. Since many of the people I wrote about hold doctoral degrees, I have not used professional titles before their names in the text. See Appendix A for full names, professional titles, and jobs at the NIH. I was told that the NIH is a "jungle of acronyms," apparently not an uncommon feature of the federal bureaucracy. So I use the full names and abbreviations of the entities interchangeably. There's another appendix defining each of the acronyms used in the book. I have saved most of my observations and opinions for the final Conclusions chapter rather than intersperse them in the body of the text; which, for the most part, has been preserved for reporting.

As in each of my books on academic medical centers, I sent to each of the people interviewed—222 from January, 2007 to November, 2008[†]—drafts of what I wrote based on what they told me for corrections, additions, and deletions. The origin of most of the quotations and facts is referenced. Some of the quotes, however, are not attributed because the speakers wished to remain anonymous. In some cases I have transformed direct quotes into text, but the source has usually been identified.

I thank Gert Brieger, Michael Gottesman, Steven Hyman, Holly Beckerman Jaffe, Jocelyn Kaiser, Ruth Kirschstein, Norka Ruiz Bravo, Alan Schechter, Harold Varmus, and Elias Zerhouni for reviewing portions of, or all of, the manuscript and giving me valuable advice.

---

[*] Which reflects my being a cardiologist.
[†] See Appendix A.

Paula Cohen in the Office of Communications and Public Liaison at the NIH arranged many of the interviews, informed me about who does what, and guided me through the multi-building campus. Barbara Harkins, archivist and librarian in the NIH Office of NIH History, helped me assemble the pictures.

At the Oxford University Press, I thank my editor Craig Allen Panner and his colleagues Molly Hildebrand and David D'Addona for their expertise and continuing encouragement and Sumitha Nithyanandan, who processed the manuscript so skillfully. The indefatigable Phyllis Farrell reviewed the proof and found, as she always does with my manuscripts, several mistakes that I missed. Feel free to blame me for those that remain.

INTRODUCTION

IN THE WASHINGTON suburb of Bethesda, Maryland, about nine miles north of the White House, is a 310-acre campus of 75 buildings in which more than 18,000 workers, one-quarter of whom hold the degrees of M.D. or Ph.D., and sometimes both, are employed by the federal government with the sole purpose of improving the health of men, women, and children through biomedical research. This is the National Institutes of Health (NIH)—"a knowledge-generating institution, not bound by local or state interests," as one of the senior scientists there described it,[1]—to which Congress allotted more than $29 billion during 2008.*

Several of the first eight years of the twenty-first century were not the most agreeable at the United States government's premier center for biomedical research and for the medical investigators and institutions throughout the country who depend on the NIH for the financial support that they need to do their work. After doubling to $26 billion from 1999 to 2003, the budget grew by less than two percent per year on average, which was less than the rate of inflation, to $29.5 billion in 2008, making funding, recruitment, and retention of investigators in the NIH laboratories and at the medical schools, teaching hospitals, and research institutes that the NIH supports, increasingly difficult.[3]

The NIH administration was also troubled by several instances in which NIH investigators bypassed the rules governing conflict of interest. The policies, unpopular among biomedical investigators, of

---

* For a discussion of the socialization of biomedical research,[2] see Richard Lewontin: "The socialization of research and the transformation of the academy," in C. Hannaway, ed., *Biomedicine in the twentieth century: Practices, policies, and politics*. Washington, DC: IOS Press, 2008; p. 9.

the federal administration limiting the use of stem cell research presented complex political problems for the leaders of the NIH.

History*

In 1798, the very young federal government created the Marine Hospital Service (MHS)† to provide medical care for merchant seamen. In the 1880s Congress charged the MHS with examining passengers arriving on ships for clinical signs of infectious diseases, especially cholera and yellow fever, to prevent epidemics. Then, in 1887, the MHS created a one-room laboratory in the Marine Hospital at Stapleton, Staten Island, New York, and authorized Joseph J. Kinyoun, a young MHS physician trained in the new bacteriological methods, to conduct studies related to the service's mission. Kinyoun called this facility a "laboratory of hygiene" in imitation of German facilities and to indicate that the laboratory's purpose was to serve the public's health. What started as this small laboratory would become, in time, the National Institutes of Health (NIH).

In 1891, the Hygienic Laboratory, as it came to be called, was transferred to Washington, D.C., and in 1904 it moved into a new building at 25th and E Streets, NW, for which Congress had allocated $35,000. This act launched a formal program of research and allowed the Service—the government had renamed the Marine Hospital Service as the Public Health and Marine Hospital Service—to hire scientists with Ph.D. degrees. Previously, only physicians had been members of the professional staff. In 1930, an Act of Congress changed the name of the Hygienic Laboratory to the National Institute (singular) of Health and authorized the establishment of fellowships for research into basic biological and medical topics.

Seven years later, Congress established the National Cancer Institute (NCI), which foreshadowed the categorical-disease structure of NIH that has characterized the agency since that time. The NCI was authorized to award grants to non-federal scientists for research on cancer, the first step into what would become the extramural program of the NIH, and to fund fellowships at NCI for young investigators. A research facility was constructed, "Building 6" of the new NIH campus in

---

* This history was adapted from the work of NIH historian Victoria A. Harden. See http://history.nih.gov/exhibits/history/index.html.
† The predecessor of the U.S. Public Health Service (PHS).

Bethesda, Maryland,* and was first occupied in 1938. In 1944 legislation, NCI was specifically designated as a component of NIH.

The proliferation of institutes, often founded through the enthusiasm of public advocates, began in 1948 with the establishment of four more categorical institutes and the renaming of the NIH to the National Institutes (plural) of Health. By 2008, the NIH consisted, organizationally, of 20 institutes, seven centers, and the office of the director.

---

* The Bethesda property, called "Tree Tops," had been the retirement home of Luke and Helen Woodward Wilson. Mr. Wilson worked in the family firm of men's clothing importing and manufacturing. His wife was the daughter of one of the founders of the Woodward and Lothrop department stores. In August, 1935, the Wilsons deeded 45 acres of the property to the government, and the National Institute of Health moved there. Designed in the Georgian revival style, the first buildings created the ambiance of a college campus. Details about the transfer of the property and the early funding of the NIH are described in "70 Acres of Science" by Michele Lyons, which can be read and downloaded from the NIH website: http://history.nih.gov/01docs/historical/documents/70AcresofSciencejuly14.pdf. 2006.

THE NATIONAL INSTITUTES OF HEALTH
1991–2008

CHAPTER ONE

Intramural Research Program[1]*

OTHER THAN FOR administrative expenses, about 85 percent of the $29 billion budget (2008) supports the work of investigators in medical schools and research institutes through the "Extramural Research Program."† About 10 percent supports those who work in the laboratories at the NIH campus in Bethesda or at one of the satellite sites. This constitutes the "Intramural Research Program."

The government created the intramural program to support research that was thought too risky and difficult to pursue elsewhere because of its unique or challenging nature or its large size. The program also provides a home for studies important to public health, such as vaccine development, and for the study of rare diseases.[2]

"It was [begun] because of the federal commitment to support basic research that developed in the immediate aftermath of World War II," said David Korn, then at the Association of American Medical Colleges. "A major role for the intramural program was to house the

---

* For the early history of the intramural program, see B.S. Park, Disease categories and scientific disciplines: Reorganizing the NIH intramural program, 1945–1960; in C. Hannaway, ed., *Biomedicine in the twentieth century: Practices, policies, and politics.* Washington, D.C.: IOS Press, 2008; pp. 27–59.
† See Chapter 2.

very best research possible and to serve as a premier training ground for new biomedical scientists."[3]

Despite the many formidable past and current accomplishments of the investigators, trainees, and staff, "the intramural program is under-appreciated now," said Eugene Braunwald of the Harvard Medical School, who worked at the NIH from 1955 to 1968. "This contrasts to when I was there. Then it had the critical mass and support needed to advance medical research with the enthusiastic support of the entire academic community. Now the non-NIH academic community often feels in competition with the intramural program for scarce resources."[4]

Herbert Pardes, who directed the National Institute of Mental Health (NIMH) from 1978 to 1984, agrees. "It was far and away the best mental health research place in the world. Today, there are many more places doing work of similar quality. Also, the recent financial cuts have hurt it."[5] Others concur.[6–8]

Careers at the NIH[1,8–20]

Why would a medical scientist prefer working at the NIH to working at a leading medical school as a faculty member? Ability to take on risky projects, "incredible"[21] resources, talented colleagues, and not having to teach or apply for grants[22] (often given as the leading advantage) are the most often-stated reasons. And then there is the prestige of being able to say, "I work at the NIH."[19]

Although the NIH is an agency of the federal government, few seem to choose to work there primarily to perform a public service.[23] Rather than say that they are federal employees, most identify themselves by their professions. They think of the NIH as a university with a campus* and tenure, and of themselves as members of a faculty.[25]†

An NIH appointment allows investigators to undertake promising projects with a higher-than-average risk of failing and to pursue the work for many years if necessary. As one of the institute directors said,

---

* In response to the attacks of September 11, 2001, security at the Bethesda campus has greatly increased. "The place has become padlocked," said one former institute director. "It used to be such a free place."[24]

† The quality of the members of the intramural program is such that, as of October, 2008, 49 are members of the National Academy of Sciences and 55 of the Institute of Medicine.[26]

"You have tremendous flexibility to chase after new ideas and to start the experiments at once instead of waiting for grants to be approved. Of course, you have to do it responsibly."[10]

NIH investigators in the intramural program say they have almost "unlimited"* resources to conduct their research. As one former NIH investigator noted, "You don't have to scrimp and save. I quickly learned to spend all my money by the end of the fiscal year so that I wouldn't lose it to the budget of someone who had overspent."[28]

To a significant degree, scientists can choose how to spend their time. If they don't wish to provide general medical care,[8] clinicians can limit their contact with patients to those with the particular diseases they are studying.[29] If they don't want to teach, they don't have to, but for those who enjoy such work, there are many opportunities to do so with trainees with different amounts of experience.[11,29]

Other than running their laboratories, there's little administrative work required of scientists at the NIH unless they choose to assume it,[1] which would often come with the senior positions at medical schools and academic medical centers for which many are recruited.

"There are more tensions at medical schools than here," said Anthony Fauci, the director of the National Institute of Allergy and Infectious Diseases. Fauci, the longest-serving institute director—he was appointed in 1984—has become a prime candidate for such positions and has rejected them all.

> I can participate in both clinical and basic research and have a broader impact on my specialty of HIV/AIDS than if I were working anywhere else. The NIH lets me continue my research and gives me a bully pulpit to advance the field in a public way. We work in a highly academic, scientifically sophisticated setting and have legions of colleagues to talk and work with. The critical mass here is different from any other place I've seen.[11]

"My years at the NIH were the most personally fulfilling time of my career," said Philip Pizzo, now dean of the Stanford University School of Medicine.[30] When Pizzo, a pediatrician, started at the NIH as a

---

* Elias Zerhouni, director of the NIH from 2002 to 2008, adds, "They wish that was true, but it is not. NIH researchers get a finite budget for their lab determined by their scientific director and cannot exceed it. The average total budget per established investigator is around $2 million including all costs. Not bad but not unlimited."[27]

trainee in 1973, he was assigned to help care for an 11-year-old boy in the Clinical Center, the NIH hospital, who had developed pancytopenia, a disease in which the blood-forming cells malfunction. Since "Ted" now had too few leukocytes, the white blood cells that fight infection, he had to live in a special room, isolated from people who might carry the bacteria that could kill him.* "We tried all the techniques we had. He lived for seven years but died of hemosiderosis," an overabundance of iron that the boy had accumulated from the transfusions he had needed, a condition that can be treated more successfully now. "His dramatic story had a big impact on me. I had seen him nearly every day."[30]

"It provides tenured investigators safe positions for the rest of their working lives," said Arnold "Bud" Relman, former editor-in-chief of the *New England Journal of Medicine*. "The doors are open to successful investigators if they want to enter academic life," some of whom leave the NIH to become deans, department chairmen, division chiefs, and institute directors at leading medical schools. Relman knows, however, that many do not want to move, as he discovered when trying to recruit Anthony Fauci to the University of Pennsylvania when Relman was chairman of the department of medicine there. "I've got everything I need at the NIH," was Fauci's unanswerable response to Relman's entreaties.[8]

Michael Lauer was a member of the cardiology department at the Cleveland Clinic[31] when Elizabeth Nabel, the recently appointed director of the National Heart, Lung, and Blood Institute (NHLBI), recruited him to direct the Division of Prevention and Population Sciences at the institute, beginning July 1, 2007. The new job would give him the opportunity to lead a more comprehensive research enterprise than was possible at the Clinic.

Moving from Cleveland to Bethesda, however, presented Lauer with problems. His family was well established in Cleveland, where his wife is a physician at the Clinic and their two teenage children attend schools they like. Since the Lauers are Orthodox Jews and need to be together and not working on the Sabbath, the NIH allows him to maintain a relationship with the Clinic where he works on his research on Friday and can then participate with his family in religious activities beginning at sundown. He flies back to Bethesda on Sunday night.

---

* "Ted" was the son of a senior NIH scientist.

Despite the traveling, Lauer is very pleased that he moved to the NIH. "I give people we're trying to recruit five reasons for coming, the same ones I have found," said Lauer.[22]

1) Superb research administration and leadership opportunities,
2) All you're expected to do is high quality research,
3) Our "think tank" where we get together and brainstorm regularly on current and future research, often with experts from the outside,
4) Program development. We plan about 20 to 25 percent of the projects, often large clinical trials, that will be performed by investigators outside the NIH,
5) Participation in professional activities, like the American Heart Association in my case, that the institute encourages us to do.

Staff Positions at the NIH[1,32-35]

There are four principal ranks available to investigators at the NIH:

- *Research and Clinical Fellows* have doctoral degrees. In 2008, there were about 3,800 fellows.*
- *Staff Scientists* and *Staff Clinicians* have completed their training. The appointments are time-limited and renewable. They do not receive independent resources and are not on the tenure track. The jobs are tied to the laboratories of principal investigators (PI), about half of whom have staff scientists working with them. If their laboratory should close and their work has been satisfactory, the NIH will try to find positions in other laboratories for them. With a few exceptions for special team projects, no more than one staff scientist can work with each PI. In 2008, there were about 1,000 staff scientists and 200 staff clinicians. These positions are analogous to research assistant professors, research associate professors, and research professors at universities.
- *Investigators* also have time-limited, renewable appointments, but, unlike staff scientists and staff clinicians, they are on the tenure track and have committed independent resources from institutes and centers. Investigators are chosen by committees after national searches. About 40 new tenure-track investigators are appointed

---

* Since the number of personnel in various positions at the NIH constantly changes, I will use approximate numbers.[36]

each year. These positions are analogous to tenure-track assistant professors at universities.
- *Senior Investigators* have tenure, usually after six to nine years on the tenure track for those already working at the NIH. Tenure at the NIH brings a long-term commitment of independent resources, including salary, operating budget, personnel, and space, for the conduct of an independent basic, clinical, or epidemiological research program in an institute or center. Candidates for these positions are nominated by search committees (if outside the NIH but not if being considered for promotion from within the NIH), are reviewed by an NIH tenure committee, and are appointed by the deputy director for intramural research. The NIH appoints about 30 senior investigators each year. About one-third are promoted from more junior positions at the NIH; the rest are recruited from other institutions. The title of senior investigator at the NIH is analogous to associate professor with tenure or professor with tenure at universities.

In 2008, the intramural program at the NIH employed about 1,140 principal investigators who conduct about 2,000 research projects. About 900 are tenured and 240 are in the tenure track.[37] Fifty to sixty percent of investigators become tenured as senior investigators after thorough internal and external reviews. Accordingly, most of the tenured investigators have been promoted from among non-tenured candidates at the NIH. A few are recruited from elsewhere directly to tenured positions, usually to direct specific programs that the institute directors want to develop, but seldom from industry.

Some of the institutes use the titles of "Lab Chief," for directors of basic research programs, or "Branch Chief," for directors of clinical research programs, to designate leaders of groups of investigators and senior investigators who work in similar research areas.[38] The scientific directors in the institutes and centers make these appointments, which are not official intramural professional designations.[32,39]

Problems about Working in the Intramural Program

Not all investigators prefer working in the intramural program to working at universities. "I never went intramural after my training there," explains Guy McKhann, former chief of neurology at Johns Hopkins University School of Medicine. "It's too secluded and isolated for me. I'm surrounded by greater depth and breadth here at Hopkins."[40] A senior branch chief, committed to the NIH because of its scientific opportunities, and with family in the area, may dream of creating and managing a larger entity, "based on meritocracy." A research-intensive academic medical center may present such an opportunity, which is seldom available at the NIH, "where resources shift more gradually,"[12] McKhann agrees.[40]

Robert Gallo, who moved to the University of Maryland in 1995 after a distinguished 30-year career at the NIH that began with a traineeship in 1965, said, "I wish I'd left ten years earlier. The NIH is a great place to start and build a career, but the advantages decrease when one's career is well established."[13] Like some others who leave the NIH, Gallo wanted to develop a more comprehensive scientific and clinical program than was possible within the NIH's bureaucratic organization. "I can now be more entrepreneurial and have been able to build an organization where it's easier to apply laboratory discoveries in my field [virology, particularly HIV/AIDS] to patient care."[13]* Practical considerations included a higher salary and a place to park, often difficult to find at the NIH. "I'd consider taking a job just for a parking space," he added, "somewhat tongue in cheek."[13]

One of the reasons the Nobel Prize–winner Arthur Kornberg left the NIH for university life in 1952 was because "I began to experience mounting irritation over the cumbersome and uninspiring NIH bureaucracy."[41]† Kornberg was not alone in complaining how working for the government can interfere with the activities of intramural scientists, saying:

- Instead of supporting excellence in research, routine administration and bureaucracy are like a straitjacket.
- Authority and power are diffuse, delaying timely decision-making. Titles do not necessarily bring the authority one would expect.
- There are too many levels of review.

---

* Gallo is director of the Institute of Human Virology at the University of Maryland School of Medicine in Baltimore.

† Another reason Kornberg left the NIH[42] was his concern that the opening of the Clinical Center in 1953 with its 500 beds and 1,000 laboratories might lead to clinician domination of scientific decisions. Kornberg later admitted that this judgment was erroneous. Arthur Kornberg died on October 26, 2007.[43]

Michael Gottesman, who directs the intramural program, agrees that "many factors tend to make our scientists' lives more complex. As an example, making renovations can be frustrating. DHHS [Department of Health and Human Services] has complex rules about this."[44] Regulations to protect human subjects, though necessary, and ethics rules make conducting clinical research more difficult. "All these checks have eroded the working lives of independent scientists. Some call it 'death by a thousand cuts'."[44]

Although the government's bureaucracy frustrates many,[40,45,46] "financing for research is easy," Gottesman reminds his colleagues, "so most of our scientists tolerate these problems. Paperwork at the NIH is much less time-consuming than writing grants, which can take up to one-third of the time of investigators at universities."[44]

Reviews

Although staff investigators at the NIH do not have to apply for grants—the bane of investigators in medical schools—their work is reviewed every four years by scientists recruited from outside the NIH. The reviews are conducted by Boards of Scientific Counselors (BSC), one for each NIH institute. Each board consists of about ten members who serve for five-year terms. Joined by ad hoc reviewers, members of the boards visit the NIH as many as three times per year to perform the reviews.*

Most intramural investigators receive satisfactory reviews. In some cases, however, the reviewers may advise that the resources of the laboratory be reduced or, occasionally, closed. The reviews can also advise increasing the size of the laboratories of successful investigators.[34]

Richard Wyatt,† deputy director of the office of intramural research, explains why the pain may be worse for investigators at medical schools than at the NIH. "It's all or nothing at medical schools since they are so dependent on getting their grants. Here, it's more a matter of degree.

---

* Intramural researchers report to the reviewers what they have *done*, while extramural investigators apply for funding with grants that describe what they *will do*.[15] Thus, the work of the intramural investigators is evaluated retrospectively, whereas the process in the extramural program is prospective.[15]

† Wyatt, a rear admiral, directed the Public Health Service Commissioned Corps at the NIH between 1987 and 2007.

Lab closures at NIH are real, but it generally takes two unfavorable reviews with reductions for that to happen."[47]

The number of laboratories closed varies among the institutes. For example, in the National Cancer Institute (NCI), the largest of the institutes with 25 percent of the intramural funds, approximately 40 of the NCI's 300 laboratories were closed in the past four years (three percent per year).[1] "With flat budgets and declining purchasing power," Michael Gottesman writes, "more laboratories have been recommended for closure in recent years. The intramural program has 100 fewer independent research programs in 2008 than it had in 2004."[34]

Although people at the NIH say that the reviews have become increasingly quantitative and critical[48]—"they hold us to R01* standards," said one branch chief[12]—critics, mostly in medical schools, question whether the reviews and the decisions by the NIH leadership adequately deal with investigators whose laboratories have become less productive.* "NIH is not so good at that [closing programs]," said one of the institute directors.[51]

Although the resources in investigators' labs can be reduced, "it's tough to fire us since we work for the government," acknowledged one of the institute directors.[48] As proof of this, no institute director has been overtly fired in recent decades.[52] In a few cases, however, NIH directors have successfully "convinced" institute directors (without officially relieving them of their positions) that they should step down, and either depart the NIH or take other positions there.

Ironically, some NIH staff members feel "trapped" in their jobs at the NIH just because they do not have to apply for grants.[12] Universities interested in recruiting mid-level and senior investigators may be concerned that those working at the NIH will not be as able to compete successfully for extramural grants; particularly those who never had to do so previously.[12] Though NIH investigators can establish their scientific reputation through publishing their research in competitive journals alone, in research-intensive medical schools, publishing without grant support can reduce the chances that faculty members will be promoted or acquire tenure.

---

* The NIH describes the research project grant, known as the R01, as "the original and historically oldest grant mechanism used by NIH. . . . R01s can be investigator initiated or can respond to a program announcement or request for application"[49] and constitute about half of all applications to the NIH.[50]

Training

The intramural program at the NIH supports several programs to train those interested in, or committed to, biomedical research.[1,15,53-56]

The programs for students and graduates considering medical or research careers, and their approximate numbers of students, are:

- Summer internship program in biomedical research for high school students, who are at least 16 years old or in college, and come to the NIH during the summer to work in one of the research programs—1,000 students.
- Post- baccalaureate ("post-bac") intramural training award for one or two years for graduates who received their bachelor's degrees less than two years previously and are deciding whether to seek medical or graduate school training—600 graduates.

Programs for medical, dental, and graduate students:

- Summer research fellowship program for medical or dental students to work with NIH investigators—about 40 to 50 students.
- HHMI/NIH Research Scholars Program, also known as the Cloister Program,* for American medical, dental, and veterinary students, jointly sponsored by the Howard Hughes Medical Institute (HHMI) and the NIH. The students live in the Mary Woodard Lasker Center for Health Research on the NIH campus. HHMI provides $27,000 a year for rent, food, and other living expenses, and the NIH provides the facilities, research laboratories, and investigators. Each member works with a member of the intramural program—42 students, usually for one year.[56]
- Year-off training program for medical and graduate students who participate for one and occasionally two years in which they take

---

* The HHMI/NIH program is based in the Cloister, built in 1922 for the Sisters of Visitation. The federal government bought the land and buildings, which were adjacent to the NIH, in 1984. When members of the program and NIH officers found the sisters' cells "less than adequate,"[56] HHMI built the Hughes House wing, which in the NIH's description of the program "provides you with a comfortable, attractive home during your stay at the NIH and enables you to be within walking distance of your laboratory." The Cloister also includes a lecture room where investigators from the NIH and other institutions present seminars each Monday evening.[56] See http://www.hhmi.org/science/cloister/financial.html for more about this program.

leave from their universities, work full-time at the NIH, and live on the campus—40 students.
- Clinical electives program for third- and fourth-year medical and dental students to work in the Clinical Center with NIH specialists—52 students.
- Clinical research training program (CRTP) for medical students to learn how to perform clinical research, a residential program that the NIH developed because most medical schools do not offer this training[57]—30 students/year.
- Graduate partnership program (GPP) for students matriculated at medical and graduate schools who perform their research in NIH laboratories but receive their course work at, and will receive their degrees from, their universities.[15] The NIH pays the university about $30,000 for tuition and miscellaneous expenses and gives the student a stipend of $20,500*—400 students.[58]
- Graduate partnership program with Cambridge and Oxford universities. These students—who have GPA scores averaging 3.9[15]—spend half their time at the universities and half at the NIH. The universities, to which the NIH pays the tuition, award the degrees. Since each student has two supervisors, one at the NIH and one at the university, the NIH believes the relationship between trainee and supervisor to be less "master/apprentice."[15] Each student in the Oxford/Cambridge program receives $3,000 per year for travel and a laptop computer configured to acquire all information on the NIH website. At the NIH the students can live in a house owned by the Foundation for Advanced Education in the Sciences—80 students.

Programs for postdoctoral investigators:

- Postdoctoral program for graduates with M.D. and Ph.D. degrees for up to five years—1,800 positions. About 800 are Americans.
- Fellows. More senior than postdoctoral trainees, for up to three years—300 in clinical research, 600 in basic research.

Except for a residency in clinical pathology, the NIH does not provide clinical training for interns and residents in the NIH's hospital. "The reason . . . is that we do not have the variety of patient exposure

---

*Claiming that the NIH provides inadequate financial support, some universities have resisted joining this program.[15]

to support these programs, and our primary mission is research, not training," explains John Gallin, director of the Clinical Center. "We do have robust fellowship programs [which medical doctors enter after internship and residency] that support our research mission. Many of the programs require partnership with other centers to assure adequate exposure to different types of patients."[59] The absence of interns and residents decreases the administrative work of these physician investigators, who also direct clinical services.[60]

Graduate Degrees

Towards the end of the 1990s (and not for the first time),[61] the NIH considered developing a graduate program that would award Ph.D. degrees in medical science.[1,15,57,61–66] With more than 1,200 principal investigators, Harold Varmus, the director of the NIH from 1993 to 1999, and Michael Gottesman, then (and now) head of the intramural program, reasoned that the NIH had the professional strength to offer the degree. An internal survey showed that about half of the scientists working at the NIH favored such a program. "We would offer tutorial sessions and less formal class work," said Gottesman, "like the Oxford and Cambridge colleges do."[57]

With so much talent at the NIH, Varmus believed that a Ph.D. program could provide superb teaching and supervision for candidates. Not having to write grants would give the staff who wanted to participate the time to prepare and deliver lectures and conduct seminars. He assured scientists at the universities that the cost of the plan would not be assigned to the extramural program.[62]

Varmus wanted the NIH to become more like such outstanding academic institutions as the University of California–San Francisco, where he had spent most of his career.[65] He saw the graduate scheme improving the intramural program by attracting outstanding candidates to the NIH, the best of whom would join the staff after receiving their degrees and postdoctoral training. The program would emphasize translational[*] research and bioinformatics and capitalize on the resources of the NIH's Library of Medicine.[65]

---

[*] Translational research applies basic discoveries to clinical medicine. "Bench-to-bedside" is the frequently employed mantra.

Gottesman presented the plan—15 students for a five-year course of study—to the NIH director's Advisory Committee,* a congressionally mandated group of about 15 leading physicians and scientists from outside the NIH.[57] Committee member Eric Kandel, the Columbia University scientist whose work on memory[67] would bring him the Nobel Prize in 2000, supported the idea. "I think our function on this committee is to make sure that NIH is as strong as possible, not that Harvard or Columbia is as strong as possible," he added in response to the accusation that an NIH program would attract superior candidates who might otherwise train at leading universities.[62]

A few members of the committee, as well as some other critics,[68] opposed the program. Cell biologist Marc Kirschner from Harvard thought that the program was inadequately developed and, in its current form, was "not a realistic proposal."[69] Shirley Tilghman, a molecular biologist who had trained at the NIH and would soon become president of Princeton University,[70] said, "we right now in this country have an excess of Ph.D.s trained in the biological sciences. For the NIH to turn around and start a graduate program sends the wrong symbolic message to the community." Tilghman had chaired a study for the National Research Council that urged universities to freeze the size of graduate programs.[62†] Others told Varmus that the institute wasn't equipped to do it, that the NIH was known for its postdoctoral programs and should continue to emphasize that work. Some believed that the federal government "shouldn't be in the degree-granting business."[57‡]

Though a majority of the committee favored developing the plan further,[57,62] Varmus decided "not to go to the mat for it,"[66] and withdrew the proposal.[62,71] As a replacement, the NIH developed the graduate partnership program,** which is "better than the degree program ever would be," said Michael Lenardo, a leader of the

---

* See http://www.nih.gov/about/director/acd/acdcharter.htm.
† "Bosh," Tilghman said to me about the fear that the NIH would steal top candidates from top schools. She also observed that the original plan for 15 students would eventually "morph into 100," further overpopulating the already excessively large market of Ph.D.s with biomedical training.[65]
‡ The Uniformed Services University of the Health Sciences and the military academies offer degrees.
** See above.

Oxford/Cambridge program at the NIH.[15] Gottesman agrees. "The universities as well as the NIH benefit from the partnership program. We ended up where we wanted to be, bringing more students to us."[57]*

Foundation for Advanced Education in the Sciences

The Foundation for Advanced Education in the Sciences (FAES) is a not-for-profit organization, governmentally separate from the NIH, that for almost 50 years has offered postgraduate courses in biomedical science.[73,74] Most are taught by NIH scientists, including fellows who learn teaching skills from participating in the courses. The Foundation does not award degrees, although the courses may be applied for credit toward degrees earned from universities.[75]

Currently, FAES offers about 200 courses to an enrollment of some 2,000 students. The average size of classes is 20 students. The courses, which may be taken by all NIH scientists, from trainees to those of emeritus status, are also open to students who are not NIH employees.

The courses are taught after normal working hours, usually starting at 5:30 P.M. in conference rooms at the NIH no longer required for routine purposes. Among the courses taught is English for non-English speakers and those for whom English is not their first language. This course is particularly popular with the many trainees who come to the NIH from other countries. As of 2008, NIH had more than 2,000 foreign fellows on its campus, representing over 90 countries, primarily China, India, Japan, Korea, and the European Union.[27] The Foundation also has courses teaching foreign languages; some of the teachers for these courses are not NIH employees.

FAES also runs a bookstore for the NIH, "which is subsidized by other FAES activities or requires subsidies from other FAES activities" says Constance Noguchi, the dean. The Foundation provides health

---

* Although he could not create a graduate school at the NIH, Varmus did so at the Memorial Sloan-Kettering Cancer Center (MSKCC), where he became president on leaving the NIH in 1999. In 2006, the first class of six students arrived to study for the Ph.D. degree in cancer biology at the Louis V. Gerstner, Jr., Graduate School of Biomedical Sciences at MSKCC.[72]

insurance programs for people working or training at the NIH who are not eligible for government benefits. FAES also owns a building that can be reserved for social and business meetings across Old Georgetown Road from the NIH.[74]

Office of NIH History

The Office of NIH History with the DeWitt Stetten, Jr., Museum of Medical Research, a program within the office of intramural research, is the comprehensive repository of historical information about the NIH from its founding in 1887.

"The history of the NIH is not that well known," said Robert Martensen, the director since October 2007, "but we're the place to go."[21]* Martinson, an emergency room and intensive care physician, bioethicist, and medical historian,[76] succeeded Victoria Harden, who founded the program in 1986.

Martensen describes several features of the office:

> To increase the visibility of NIH History, the History Office, in collaboration with various Institutes, sponsors one- or two-year Stetten Postdoctoral Fellowships for historians of science and medicine. Each develops a project on some aspect of biomedical research or research policy that NIH has sponsored. For 2008–2009, the Office will house four Stetten Fellows. In addition, it has started a Summer Intern program for college students with demonstrated interest in the history of science, technology, and medicine. The Office sponsors a public lecture series and conferences. The Stetten Museum has no home of its own, but mounts displays on the internet and in different buildings on the Bethesda campus.[21]

Since the history office has a "tiny budget,"[21] it funds many of its programs through collaborations with other components of NIH, such as the National Cancer Institute and the History of Medicine Division of the National Library of Medicine.[21]

---

* See the History Office's website, http://www.history.nih.gov/, for more historical information about the NIH.

Commissioned Corps; Associate Training Program

Three hundred sixty-eight physicians, dentists, veterinarians, nurses, dieticians, and pharmacists who work at the NIH—about six percent of the professional employees there—are members of the all-officer Commissioned Corps of the United States Public Health Service (PHS),[*] which is one of the seven uniformed services[†] of the United States.[77,79 81]

The Commissioned Corps at the NIH enjoyed its greatest attraction to talented young physicians when the "doctor draft" was in effect.[‡] In addition to the traditional uniformed services, doctors could serve in the Public Health Service Commissioned Corps, and one of the options available to the most competitive of them was the NIH's Associate Training Program, which started in 1953.[83]

Associates were assigned as clinical, research, or staff associates. Clinical associates worked primarily in the Clinical Center, taking care of patients and participating in clinical research under protocols established by the NIH research institutes. Research associates learned and performed laboratory research. Staff associates received training to become scientist-administrators.[83]

Each associate in the Commissioned Corps entered the program with the Navy-equivalent rank of (full) lieutenant[47] or lieutenant commander and received tax-free rental and subsistence allowances in addition to base pay. Other benefits included medical and dental care and access to military post exchange (PX) facilities.[83]

---

[*] The NIH was part of the Public Health Service until 1953, when it became an independent agency of the newly created Department of Health, Education and Welfare—renamed the Department of Health and Human Services in 1980. The Public Health Service is also an agency of the Department of Health and Human Services. Some of the 6,000 members[77] of the Commissioned Corps also work at the Agency for Healthcare Research and Quality, the Food and Drug Administration, and the Indian Health Service.[78]

[†] The Army, Navy, Marine Corps, Air Force, Coast Guard, National Oceanic and Atmospheric Administration Commissioned Corps, and Public Health Service.

[‡] The Doctor's Draft Law from 1950 to 1973 obligated physicians and dentists for two years in one of the uniformed services. When it ended, pressure on the best and the brightest to train at the NIH decreased. As one longtime NIH investigator observed, "the NIH never fully recovered" from the loss of these outstanding young medical investigators and clinicians.[82]

Other opportunities than the NIH were available to doctors to fulfill their service obligations in the Public Health Service, such as the Communicable Disease Center, widely known as "the CDC" (now the Centers for Disease Control and Prevention). The CDC appealed to applicants with particular interest in infectious diseases. Those commissioned into the Public Health Service who were unsuccessful in securing a position at the NIH or CDC could be assigned to PHS hospitals, Indian Health Service hospitals, or the Arctic Health Research Center, among other places.

At its peak, about 1,000 of the professionals at the NIH were members of the Commissioned Corps.[77] In addition to the stimulus provided by the doctor draft, the financial provisions of a position in the Commissioned Corps exceeded those provided at an equivalent rank in the federal Civil Service. When the provisions of Civil Service employment at the NIH improved, many of those who had advanced to permanent positions at the NIH sought the Civil Service instead of the Corps. Many fewer who accept physician appointments at the NIH now are members of the Commissioned Corps.[47]

Officers in the Commissioned Corps receive base pay and possibly some professional special pay, which is fully taxable, and housing and food allowances and moving expenses that are not taxable. An accession bonus of up to $30,000 may be awarded to a new dentist, nurse, or pharmacist officer who agrees to stay for four years. These are one-time payments often used to recruit medical specialists with high earning power in civilian practice. Compared with the Civil Service, the Corps offers better vacation and sick-leave provisions, use of the exchange and commissary, and space-available seating on military airplanes. Health care is free for the Corps officer and his or her immediate family. However, compared with the relatively routine promotions in the Civil Service, promotions in the Corps are fairly competitive, especially at the higher ranks.

Most professional employees at the NIH are now members of the Civil Service even though some benefits favor membership in the Corps. Non–U.S. citizens cannot be members of the Corps. Members of the Corps can be temporarily assigned for emergency duty, such as occurred with Hurricane Katrina in 2005.

Admirers of the Commissioned Corps at the NIH say that many new appointees at the NIH do not know about the advantages of being a member of the Corps compared with the Civil Service.[77,84]

Yellow Berets

Although the Public Health Service describes the Commissioned Corps, which has about 6,000 members now,[47] as a "*mobile**" health force of professional manpower," the furthest most associates during the "doctor draft" traveled from the NIH was to attend clinical rounds and conferences at the Bethesda Naval Hospital across Rockville Pike (Wisconsin Avenue) from where they worked.[83] Their association with the military was often but dimly recognizable. Most wore lab coats at work, and many never owned a uniform.[†]

During the Vietnam war they became known as "Yellow Berets" who were fighting the "battle of Bethesda."[83,85] "Although no one is certain exactly when the phrase 'yellow berets' originated, the term may have initially carried a pejorative connotation, although by the end of the war period most Associates used it as a badge of pride," writes Melissa Klein.[‡] The atmosphere at the NIH during the 1950s and 1960s arouses the nostalgic memories of many who subsequently became leaders in scientific medicine in medical schools as well as at the NIH.[4,7,17]

Many fellows came to the NIH from an internship or residency at a teaching hospital. John Potts remembers that moving from an internship at the Massachusetts General Hospital to the NIH increased his salary from $600 per year to $9,000.[87**]

Abner Notkins, who had received his M.D. at New York University and served an internship and residency at the Johns Hopkins Hospital, was appointed to one of the coveted research associate posts at the National Cancer Institute, to start in 1960. He remembers, "My supervisor, Maurice Lande, took me into a room of about 250 square feet and said, 'This is your lab, there is your lab assistant, and over there are the

---

[*] My italics.

[†] C. Everett "Chick" Koop, the Surgeon General from 1981 to 1989, had little success with the associates when he tried to enforce the wearing of uniforms. Koop always wore his uniform of a two-star rear admiral at work, in public, and when photographed.[47]

[‡] "Yellow," the traditional color for cowardice, as in "he's yellow"; "berets," an ironic allusion to the green hat worn by soldiers in the Special Forces, which the Army describes as consisting of its "most specialized experts in Unconventional Warfare."[86] See two informative articles about the Yellow Berets by Melissa Klein in the NIH website at http://history.nih.gov/articles/YellowBerets.pdf and by Buhm Soon Park in *Perspectives in Biology and Medicine* 2003;46: 334–336.[85]

[**] Potts spent eight years in the intramural program before returning to the MGH as chief of endocrinology and, subsequently, chairman of the department of medicine there.

catalogues for you to order anything you need. The library is across the hall.' The only instruction he gave me was 'do something interesting.' "[17] Notkins followed his mentor's instructions and began a career at the NIH that was to continue for more than 45 years and bring him recognition for the breadth and quality of his work.* Although getting started as an associate at the NIH is more structured these days, it could be quite an adventure in the days of the yellow berets.

Eugene Braunwald, also an NYU medical school graduate, remembers with deep appreciation the 13 highly productive years that he spent in the intramural program of the National Heart Institute (now the National Heart, Lung, and Blood Institute) from 1955 to 1968. "I'm very partial to the NIH," he said, "because it gave me so much."[4]

"Lots of doors were opening in cardiovascular physiology in the 1950s and '60s," Braunwald recalls. "Organized research programs, which we now take for granted in medical schools, were already in full bloom at the NIH then. Few other places, at the time"— Braunwald mentions Rockefeller University—"were professionalizing biomedical research as a serious, full-time occupation. Elsewhere, there still was much dilettantism."[4]

The Clinical Center had opened three years before Braunwald arrived, and the intramural program was growing rapidly. "Six to seven hundred would apply each year to be clinical associates at the NIH, so we got to choose the best of each year. You put people like that together, and things happen."[4]

Braunwald describes the leading feature of training at that time. "The experience in clinical care and research at the NIH then was unparalleled in its depth, but it wasn't broad, and it wasn't meant to be. We were there to conduct research, unlike what existed at most medical schools and teaching hospitals." The illnesses of patients at the Clinical Center could be so specialized that many associates found it necessary to take another year or two of clinical training at a teaching hospital after their experience at the NIH.† "Patients with myocardial

---

* Notkins continues his research as chief of the experimental medicine section of the Institute of Dental and Craniofacial Research. From 1984 to 1992, he was scientific director of that institute's intramural research program.
† Braunwald interrupted his NIH experience to take a senior residency in medicine at the Johns Hopkins Hospital. Upon leaving the NIH, Braunwald would become the founding chairman of the department of medicine at the University of California–San Diego and then chairman at Harvard's Brigham and Women's and Beth Israel hospitals in Boston. In addition to his other many activities, he edits leading textbooks in cardiology and internal medicine.

infarctions went to the community hospitals, seldom to us," Braunwald remembers.[4] "The same continues to be true," says Henry Masur, chief of the department of critical care medicine at the NIH.[29]

Michael Brown, a clinical associate and postdoctoral fellow from 1968 to 1971, writes:[88]

> The experience was particularly enlightening for young physicians like Joe Goldstein and me.[*] We had just completed our residencies, so we knew something about clinical medicine. We had no conception of basic research, but we were eager to learn. Joe worked with Marshall Nirenberg, who received the Nobel Prize for cracking the genetic code. I worked with Earl Stadtman, who received the National Medal of Science for opening the field of enzyme regulation. Both of us worked in simple bacterial systems. As novices in the lab we worked alongside other young physician-scientists who had preceded us into the laboratories by one or two years. The excitement was electric. As physicians, our lives had been filled with complexity and dominated by unknowns. We made clinical decisions based on inductive logic. Doctors had tried something, and it seemed to work, but no one had any idea why. Entering the pristine, logical world of pure basic science was liberating. We learned to design simple experiments that gave precise answers. While working separately at NIH, both Joe and I succeeded in drawing back the veil of mother nature and observing a simple, fundamental truth. Our lives were changed forever. We dared to dream that somehow we could reduce a complex disease to the type of rational examination that we had learned when working with simple systems. This experience launched us on our path to unravel a devastating human disease, familial hypercholesterolemia.
>
> Our eye-opening experience at NIH was shared by many young physicians of our generation. Five of us, all contemporaries in the late 1960s, went on to receive Nobel Prizes. Strikingly, three of these prizes were awarded for advances in the most devastating human diseases—Joe and I worked on coronary heart disease,

---

[*] Michael Brown and Joseph Goldstein, who have jointly directed a continuously productive research laboratory at the University of Texas Southwestern Medical Center in Dallas for more than 30 years, won the Nobel Prize in Physiology or Medicine in 1985.

Harold Varmus solved a major problem in cancer, and Stanley Prusiner made a monumental advance in understanding neurodegeneration. The other two prizes also related directly to human disease. Alfred Gilman discovered a major mechanism by which drugs and hormones alter the functions of the human body. Ferid Murad found a chemical that controls blood flow and blood pressure. In addition to these contemporaries, three other NIH trainees, Richard Axel, J. Michael Bishop, and Daniel Nathans, would receive Nobel Prizes. Remarkably, all nine of these Nobel laureates are M.D.s.

We were products of the unique juxtaposition of eager young physicians and brilliant, older basic scientists at the National Institutes of Health, and humanity was the beneficiary. My deepest regret is that those days are gone forever.

Many other leaders of specialty divisions and departments in medical schools, research institutes, and industry, and at least one university president, began their research careers in the Associate Program. "It was one of only a very few places then to get the depth of training one needed for a successful career in medical research," said Robert Gallo, co-discoverer of the virus causing AIDS.[13]

Samuel Broder, a longtime NIH scientist who started as a clinical associate in 1972 and directed the National Cancer Institute from 1989 to 1995, remembers, "at one time the best and most cutting-edge clinical research in cancer was being done at NIH."[89]

"It gave me the chance to do research full-time and get off the treadmill of training," remembered Guy McKhann, former chief of neurology at Hopkins. "It was a very open, free-wheeling academic experience with very bright people as mentors and co-fellows. It changed careers for many people."[40]

While a fellow at the NIH in the mid-1970s, Shirley Tilghman, the president of Princeton University, "felt that I'd died and gone to heaven. Someone always knew what I needed to know. It was like a sandbox is to a kid."[65] The young Shirley Tilghman might not have been given the opportunity of training at the NIH if she had applied during the doctor draft. "The associate program during the war clearly discriminated against women," remembers Harvey Klein, chief of the department of transfusion medicine and a former yellow beret. "Several senior NIH researchers at the time indicated that they would not hire a female associate because that would mean that a male physician would likely be sent to Vietnam. That policy probably influenced subsequent academic appointments for decades, since time at NIH was often like a prep school for academia."[90]*

Klein and other former trainees, such as Anthony Fauci, director of the National Institute of Allergy and Infectious Diseases; John Gallin, director of the Clinical Center; Richard Hodes, director of the National Institute on Aging; and Michael Gottesman, director of the intramural program, never left the NIH, although none of them remain members of the Commissioned Corps.

Donald Fredrickson, James Wyngaarden, Bernadine Healy, and Harold Varmus, all former NIH trainees, subsequently became directors of the NIH.[1,47]

Members of the intramural program who have won the Nobel Prize for work performed at the NIH include: Christian Anfinsen, Julius Axelrod, Carleton Gajdusek, Marshall Nirenberg, and Martin Rodbell. As of 2007, twenty scientists who were at one time associated with the NIH have won Nobel Prizes.[1] The NIH extramural program supported many other scientists who won the prize in medicine or physiology or in chemistry since the 1950s.[1,47]

The Intramural Program Today

Gradually, the attractiveness of training at the NIH for young physicians decreased. During the halcyon days of the 1950s and 1960s, there were few opportunities, even in leading medical schools and their teaching hospitals, for young doctors to receive the quality of research training offered at the NIH. As medical schools developed additional and more sophisticated research programs, often led by former associates at the NIH, the uniqueness for potential investigators of spending time there declined to the extent that at leading research institutions, few of those seeking advanced research training favored the NIH.[7,91-93] The termination of the doctor draft in 1973 removed some of the pressure on the best and the brightest to seek positions as NIH associates.[94]

Remembering when he was a yellow beret at the beginning of a 40-year career at the NIH, Phillip Gorden, director of the National

---

* Zerhouni comments: "More recently, gender inequalities at NIH have been reduced somewhat. More women and minorities have been appointed as directors or in high positions at the NIH in the past few years than ever before. For example, of the 13 directors I appointed, six are women."[27]

Institute of Diabetes and Digestive and Kidney Diseases (NIDDK) from 1986 to 1999, believes that the mentoring of the young trainees is now less intense than when the NIH was smaller. "Everyone's more on their own compared with when I was a trainee and young investigator," said Gorden.[95] "The increasing rigidity of the NIH tenure system that requires showing independence in the early stages of one's career interferes with collaboration. Young people have to separate from their mentors too soon."[95]

It was from the group of highly qualified trainees that the NIH formerly developed many members of its staff of clinical investigators. Though the salaries for trainees are competitive today, the training fellowships are much less sought-after by young American medical scientists than in the past.* As a result, the positions are increasingly filled by doctors and scientists from outside the United States where opportunities for training at comparable institutions do not exist.[95]

Many believe that the attractiveness of research careers for doctors in general has dropped.[30,91] The greater financial rewards of practice, particularly in certain specialties, and the insecurity and labor of having to apply continuously for support, often from the NIH, have soured many potentially productive medical investigators on such a career.[83,91] Philip Pizzo left after 23 years at the NIH to become chairman of the department of pediatrics at the Harvard Medical School and the Children's Hospital of Boston, where he had trained—he would later become dean at Stanford. "Fewer pediatricians were going into research. I had to return to a medical school to have any effect on this. Frankly, with my family's expenses rising, I couldn't afford to stay at the NIH."[30]

This change in career for the most able medical graduates may extend to the level of chairmen of clinical departments such as medicine. At one time, attaining such a position constituted the peak of a career in academic medicine. Now these jobs hold less allure. Dealing with the administrative and financial problems of leading clinical services and trying to recruit and sustain talented faculty with increasingly fewer resources have made clinical chairmanships much less

---

* When he was chairman of the department of medicine at the highly competitive medical school at the University of California, San Francisco, Lee Goldman "can't remember one resident wanting to go to the NIH for training recently, or any faculty member who was successfully recruited to the intramural program."[91] Goldman is now dean of the Columbia University College of Physicians and Surgeons.

attractive. Anthony Fauci, who has been offered and has continuously rejected department chairmanships as well as other senior positions at leading medical schools, has said, "Hardly anybody really wants to be a chairman of medicine any more."[11]

Some investigators both inside and outside the NIH believe that the intramural program and the morale in some members of the program aren't as strong as they used to be.[7,13,24,28,30,66,68,82,87,95–104]* When Harold Varmus started as director, he was heard to say that research in the intramural program was "spotty."

As requested by Congress,[106] Varmus formed a committee of eminent biomedical investigators in 1993 to evaluate what should be done about the state of the science and research being conducted in the intramural program.[†] Gail Cassell, the co-chair, believes that among the more important recommendations were to:[106]

- Conduct more rigorous reviews of research programs and reviews of senior scientists and scientific directors.
- Exercise more care in awarding tenure.
- Advertise widely when recruiting for open positions.
- Improve the intramural training program.
- Establish an annual perspective planning process by each institute and center to determine the allocation of resources to the intramural and extramural programs with extensive consultation with the extramural community.
- Cap the total intramural research program budget for institutes, centers, and divisions so as not to exceed the rate of 11.3 percent of the total NIH budget.

Other academics who have questioned the quality of the intramural program include Edward Benz, president of the Dana-Farber Cancer

---

* See "Is NIH's crown jewel losing luster" in *Science* 1993;261:1120–1127, for a comprehensive discussion of this topic.[105]

† Members of the committee were (affiliations at the time): Gail Cassell, University of Alabama at Birmingham, co-chair; Paul Marks, Memorial Sloan-Kettering Cancer Center, co-chair; Michael Brown, University of Texas Southwestern Medical Center, Dallas, Nobel laureate; Gerald Fischbach, Harvard Medical School; Elizabeth Neufeld, UCLA School of Medicine; Arthur Rubenstein, University of Chicago Pritzker School of Medicine; Kenneth Shine, Institute of Medicine; Maxine Singer, Carnegie Institution of Washington; Roy Vagelos, Merck and Co.; and James Wyche, Brown University.

Center in Boston and former chairman of the department of medicine at Johns Hopkins. "The extramural community generally regards the intramural research program as uneven in its quality," he said.[97] While some see the problem due to inbreeding[24,104] and lack of turnover,[104] most primarily blame the flat budgets.

The intramural program is having more difficulty than in the past in recruiting leading physician-investigators to senior positions who will stay,[9,24,60,68,102,104,106–111] Francis Collins, until recently the director of the Genome Institute,* being a notable exception. Just how many investigators have rejected opportunities to work at the NIH because of the conflict-of-interest controversy,† however, is difficult to determine.[107] A longtime clinical investigator blames the stringent regulations about conflict of interest which, he believes, "can devastate a program. We can't recruit top-quality people in clinical research and have trouble retaining the clinical people we have."

Competition for talent helps explain why many institute directors and scientific and clinical directors are promoted from within the NIH.‡ The expense of living and educating children in the Washington area and the restrictions on salaries for some specialists also contribute to the difficulty of persuading top clinician-scientists to come to the NIH.[27] Recruiting basic scientists, many of whom hold the Ph.D. degree, is somewhat less difficult since the differences in salary between the NIH and other employers is less for them than for clinicians.[95]

Medical schools and research centers, particularly for senior positions, can offer investigators higher salaries and retirement benefits and the right to pursue, and be compensated for, outside activities.[60,108] For investigators who choose to work with industry, expensive equipment usually comes with the job. Formerly, only the NIH could afford to buy such items.[98]

However, unlike in most academic centers where full-time physicians must practice within the practice plan of their medical school, NIH clinicians may have a private practice "on the side."[27] NIH

---

* And since 2009, director of the NIH.
† See Chapter 8, where the conflict-of-interest issues are discussed.
‡ Although this is what I was told by many interviewees, Elias Zerhouni says, "Not really true. During my tenure [director of the NIH from 2002–2008], I recruited 13 directors and one senior OD [Office of the Director] science position. Of the 14, 10 were not from NIH."[27]

scientists may also accept lecture fees as long as they do not come from industrial sources.[27]

Except for the most accomplished scientists, the flat budgets have reduced the number of academic jobs available at medical schools. Ironically, these financial problems may be helping to retain investigators in the intramural program, as those offered jobs elsewhere contemplate the angst of scientists outside the NIH who are trying to fund their work through the straitened extramural program.[112]

Nevertheless, the intramural program has not escaped the effects of the flat budgets. NIH has had to reduce the number of laboratories run by principal investigators. Many can stay at the NIH if they wish, but they may have to join someone else's lab or assume administrative responsibilities.[113] Paul Sieving, the Eye Institute director, said, "We lack that middle group of people, particularly physician-scientists, to succeed us."[111]

The NIH director was paid only $191,000 in 2008.[27] "It's illogical," says Edward Oldfield, former chief of neurosurgery at the NIH. "It's almost embarrassing. Such salaries are 'off the graph' for surgical subspecialties."[60]

In 2000, academic psychiatrist Dennis Charney was recruited to the National Institute of Mental Health from Yale, where he was deputy chair of academic and scientific affairs. Charney's position as chief of the mood and anxiety disorder research program and the experimental therapeutics and pathophysiology branch was a fine opportunity. The NIH would provide all the resources he needed for his research, and he wouldn't have to write grants. "It was heaven," Charney said.[114]

By 2004, however, he had left. "Dollars were the principal reason," Charney explains.[114] He found that his salary and the policy that prevented his earning money from opportunities outside the NIH provided inadequate financial resources for, among other obligations, educating his five children. So he accepted a position at the Mount Sinai School of Medicine as dean of research. In March, 2007, he became the school's dean.[114]

Joseph Goldstein from the University of Texas Southwestern Medical School in Dallas, who started at the NIH as a clinical associate in the mid-1960s and won the Nobel Prize in 1985, believes that the NIH has become "less successful on a scientist-to-scientist basis than smaller institutions like the Rockefeller[115] and Cal Tech. Size makes a difference."[100]

David Baltimore, the former president of the California Institute of Technology (1997–2006) who also won the Nobel Prize, describes the intramural program as "a haven for odd ducks who only want to do research."⁹⁶*

Steven Hyman, a neurobiologist who led the National Institute of Mental Health in the late 1990s, found that "an awful lot of intramural science is routine. High-risk stuff is uncommon."¹⁰¹ He believes that "too much security can be a bad thing. One reason that American research has prospered so much is because there is real competition based on peer review and no security."

Observers ask what is the role of the intramural program today.⁷ Hyman suspects that, "if we didn't have the intramural program now, we wouldn't create one."¹⁰¹ Some agree.¹³,⁹⁶ "Then, it was a necessity; not now," says Robert Gallo.¹³

Others disagree, believing that the research in the intramural program is better now than it has ever been and that working at the NIH is a unique opportunity. Michael Gottesman, the director of the intramural research program, writes:³⁴

> We still do a disproportionate share of the basic science research in the country, respond immediately and with superior expertise to public health problems, and continue to do paradigm-shifting research because of the ease with which high-risk studies can be undertaken and research directions can turn on a dime. As clinical research becomes harder to support on the outside, the Clinical Center remains a major resource for long-term natural history studies, rare disease research—first in human clinical trials—and remains a neutral, bias-free testing ground for new therapies.

Others at the NIH said:

- "The quality of the research has consistently improved. Our standards for recruiting and awarding tenure have become more rigorous,† and we're less bureaucratic than we used to be."¹⁵
- "It's still a wonderful place for young people to train."²⁸

---

* Among the phrases used for the acronym NIH is "Nerds in Heaven."¹¹⁶
† In some institutes, candidates for tenure must describe and defend the quality of their work before a special committee. "Like defending your thesis," said one investigator with a Ph.D.²⁸

- "There's such a great atmosphere. Everybody is interested in what they're doing, and most work extraordinarily hard. It's a place you want to come to in the morning."[107]
- "The opportunities to establish collaborations in both basic and clinical research are outstanding and probably unmatched anywhere else in the world."[107]
- "It's intellectually stimulating every day."[79]
- "You're surrounded by very interesting and intelligent people who want to do good science without the problems you meet in other jobs."[18]
- "We have access to first-rate collaborators and great informatics."[29]
- "Collaborative arrangements are very open and easy."[95]
- 'There's a special feeling about the NIH, felt at all levels. Most of us feel we work for the NIH, not really for the government."
- "I don't agree that if the intramural program didn't exist, we wouldn't create it. It's very important for science, and it helps to retain the support of Congress."[110]

CHAPTER TWO

Extramural Research Program

MORE THAN 80 percent of the NIH budget is spent by investigators working outside the NIH, mostly in medical schools and research institutes. This is the "Extramural Research Program," which, by currently supporting more than 40,000 grants for research and training,* constitutes "the mainstay of the federal funding for biomedical research at academic institutions in the U.S."[1]

"Rescuing the NIH"

In the April, 2006 issue of the *Journal of Clinical Investigation*, editor-in-chief Andrew Marks, an academic physician and investigator with strong opinions,† published an editorial titled, "Rescuing the NIH before it is too late." Marks wrote about "the utter lack of support for

---

* See http://grants1.nih.gov/training/extramural.htm, the website on extramural training opportunities supported by the NIH.
† Marks, a member of the National Academy of Sciences, is chairman of the department of physiology and cellular biophysics at the Columbia University College of Physicians and Surgeons. One of his former teachers describes him as "irrepressible and at times outrageous."[2]

biomedical science from the White House and Congress.... The White House under George W. Bush is targeting the NIH for destruction."³*

Like other basic scientists worried about whether their support from the NIH would continue,² Marks criticized the NIH for wasting "hundreds of millions of dollars... on poorly designed clinical studies whose results are suspect. The NIH should not fund large clinical studies that divert hundreds of millions of dollars away from hypothesis-driven scientific research; pharmaceutical companies should." He advised that the Roadmap for Medical Research,† NIH director Elias Zerhouni's strategic plan for the Institutes, should be "shelved and the funds restored to the pool of resources that support investigator-initiated individual R01 grants."³

Writing this paper "is something I wanted to do for a while," Marks said. "So at 4:00 a.m. one night in San Francisco, with my head still was on Eastern time, I wrote it. It was something that needed to be said."⁵

Many disagreed with Marks's opinions.⁶⁻⁸ The most intense objection came in a letter signed by the 27 institute and center directors at the NIH.⁹ The institute directors considered the Marks's attack "unfair" and "inappropriate" and praised Zerhouni for pursuing "a forward-looking approach to sustain[ing] our commitment to basic research while developing innovative ways to translate basic discoveries into clinical practice."⁹ The directors defended the value of the Roadmap and emphasized that the new undertakings represented only "1.2% of the total FY06 NIH budget,"⁹ a figure that Marks believes understates the commitment.¹⁰ The directors "strongly disagree[d] with the premise that

---

* Marks is not the only person who blames the Bush administration for the parlous state of NIH funding. "Bush has taken any number of ways to decrease funding for science," said Victoria Harden, the former NIH historian.⁴ Elias Zerhouni sees things differently. "Throughout this book there is the idea that the doubling occurred under the Clinton administration and that "bad" Bush came in and destroyed it all. This is simply not the truth. The doubling started in 1999, the year Harold left, and ended in 2003, a year after I came in. So three years of the doubling were under Bush, who personally had committed to finishing the doubling and did, against much opposition, form his own office of budget and some in Congress. In fact, they tried to rescind the 2002 and 2003 budget commitments, and after a personal appeal from me, the administration, the budget was changed back to full doubling through his personal intervention. Note that throughout the Clinton years he never proposed a doubled budget; Congress through Porter and Specter did. During the three Bush years of the doubling the reverse occurred. The administration put in the full amount."¹

† See Chapter 8.

clinical trials should only be conducted by pharmaceutical companies," arguing that "the NIH is funded by taxpayers to whom we have the responsibility of providing new information ... including the fruits of clinical trials that industry will not support."[9]

Marks is not alone, however, in complaining about the decrease in extramural funding by the NIH.[2,11-13] Critics protest that the NIH funds too many NIH-supported clinical studies and NIH-directed research proposals and contracts in the extramural program and not enough investigator-initiated proposals, like R01 grants.[13] "We don't need to direct science from the center," said Gerald Weissmann, editor-in-chief of *The FASEB Journal* and director of the Biotechnology Study Center at New York University School of Medicine.[13*] "The function of a federal agency that funds science is to respond to the innovative, novel, and exciting ideas generated by individual scientists. Considering the enormous lassitude of large organizations, I tend to doubt that we will get more bang from the buck from big science."[15] This tension between outside and inside planning is long-standing.[16]

Norka Ruiz Bravo,[†] director of the office of extramural research at the NIH,[‡] explains that there are several reasons, other than less money—83% to 84% of the NIH budget remains in the extramural program—to explain why the NIH seems to be funding fewer grants than previously:[18,19]

- Investigators are writing more grants, often "prophylactically,"[11] to reduce the likelihood of losing support.
- Since the expense of conducting research continues to rise, applicants ask for more money in their grants at a time when the

---

[*] Weissmann quotes his mentor Lewis Thomas in support of his point of view: "What [research] needs is for the air to be made right. If you want a bee to make honey, you do not issue protocols on solar navigation or carbohydrate chemistry, you put him together with other bees ... and you do what you can to arrange the general environment around the hive. If the air is right, the science will come in its own season, like pure honey."[14]

[†] Ruiz Bravo, who was born in Peru, received her doctorate in biology from Yale and did research in cell biology in Houston at the M.D. Anderson Hospital and the Baylor College of Medicine before moving to the NIH in 1990. "My first name comes from a Russian ballerina," she said. "Did you know that one of Pavlov's dogs was named Norka?"[17]

[‡] In the fall of 2008 after I had interviewed her, Ruiz Bravo's title changed from Deputy Director for Intramural Research to Special Advisor to the Director.

total amount of money in the NIH budget, after accounting for inflation, is falling.
- The number of applicants for NIH grants continues to grow despite rumors that medical research is losing its popularity as a career.
- The average success rates for all institutes in 2007 are in fact much higher than the low rates that have been widely quoted.*
- The success rate is not equal among the institutes, and frustrated applicants may quote the rates of funding in institutes whose pay-line is more or less generous. Each institute manages its own budget and has its own constituencies and preferences. However, "earmarking" for favorite projects, according to NIH officials, is infrequent.[18] The institutes that favor large center grants, which, because they can consume large amounts of money, have been called "insatiable," reduce the funds available for investigator-originated grants like R01s.[18]
- To the claim that the NIH is sponsoring clinical studies at the expense of basic research, each of these subjects is receiving the same fraction of the total NIH budget as in the recent past.

But the pain, Ruiz Bravo recognizes, is real.[21] Performing medical science is increasingly complex, and there are disciplines such as statistics that clinical investigators, in particular, must master. Doctoral candidates take longer to get their degrees, and investigators with M.D. and Ph.D. degrees spend more years in postdoctoral training than previously, sometimes taking more than one fellowship until a suitable job becomes available.[22] Physician-investigators, who have become increasingly stressed by the pressures of clinical work, postpone the time when they can apply for support, often in the form of R01 grants, from the NIH. Half of the physician-investigators who obtain their first R01 fail to obtain a second, often because they do not submit a competitive renewal.[23]

As more investigators apply for fewer funds, selecting the deserving grants from so many worthy applications has become increasingly difficult.[24] This conflict of too many R01 applications chasing too few dollars has forced the NIH to engage in what has become known as the "wait in line policy."[25] Until 2009, less than 10 percent of initial R01 attempts, labeled "A0" applications, got funded. Investigators

---

*In 2007, the NIH received 27,332 applications for R01 equivalent grants. In all, 6,462 were awarded, for a success rate of 23.6%. These applications were submitted by 21,689 investigators, of whom 6,085 (28.1%) received funding.[20]

could then submit a first revision, "A1," and, if that failed, a second and final revision, called "A2." This system changed in 2009. Now, investigators may submit only the original "A0" application and one revision, "A1."

Many receiving money through the extramural program wish that the intramural program would close and the dollars saved be added to the extramural program.[26]

Clinical Research Versus Basic Research

The modern NIH owes its development in important measure to the influence of Vannevar Bush, director of the Office of Scientific Research and Development during World War II, who persuaded President Harry Truman in 1945 to endorse a major investment of federal funds in science. Bush, an electrical engineer and science administrator, "strongly recommended," according to a recently published article, "that federal grants to universities should only support basic sciences and that the funding agency not have its own research program."[27] If this advice had been followed, clinician-investigators working at universities would not be eligible for NIH support, and the intramural program with its more than 1,000 scientists on the Bethesda campus would not exist. Although Bush's advice in this matter was not followed, controversy has persisted about whether the NIH should favor basic or clinical research.

As a basic investigator, Andrew Marks argues that the Roadmap has decreased support for his kind of science by favoring clinical research in which humans are the principal subjects.[3] Zerhouni disagrees. "The NIH should support a full spectrum of research from basic to applied including non–disease-specific sciences," he says. "Most still think linearly about research, from basic biology to applied clinical, but my experience tells me that progress comes [from] closer interactions at the interface of both biological, physical and applied sciences and from both directions."[1]

Many believe, however, that the NIH has traditionally preferred funding basic research and should increase its support of clinical research.[6,11,28] "The NIH has a very unbalanced research portfolio," said Robert Califf, a leading clinical investigator at Duke University. "Many human investigators joke that it should be called the 'National Institutes of Mouse Health,' not human health."[6]

Califf is not alone in believing that the NIH favors funding basic science and supporting Ph.D. scientists over physicians.[29,30] Former

National Cancer Institute (NCI) director Vincent DeVita has seen clinical investigators "try to present phase I trials [in which they study the chemical and metabolic effects of a new drug in a small group of human volunteers] as basic research to help getting funded."[29]

Clinical investigators believe that they have more difficulty obtaining NIH grants than their colleagues who do basic science.[31] They are right. A recent study of applicants for R01 grants by the Association of American Medical Colleges supports this impression.[32] The report showed that applicants with Ph.D. or combined M.D. and Ph.D. degrees, who by virtue of their training favor basic research, were more likely to obtain initial NIH funding than those with only M.D., degrees whose studies are more apt to emphasize clinical research. M.D.s who propose basic projects with their first application are more often funded than M.D.s applying for the first time to perform clinical studies.

Applicants with Ph.D. or M.D./Ph.D. degrees are more successful getting re-funded than those with only M.D., degrees. M.D.s are also less likely to apply for second grants. Whereas the number of applicants with Ph.D. and M.D./Ph.D. degrees has grown, the number of applicants with M.D., degrees did not substantially increase from 1983 to 2004, despite the advances in investigative medicine during this period and the doubling of overall funding from 1998 to 2003.[32]

The value of basic research to clinical research, however, should not be minimized. "We mustn't forget that fundamental research is where good ideas for clinical research often start," said Kenneth Shine, executive vice-chancellor for health affairs at the University of Texas and former president of the Institute of Medicine. "Developing the polio vaccine depended upon the basic work of Enders and colleagues showing that viruses could be grown in tissue culture. Developing a vaccine to prevent HIV infection depends as much on advances in basic research as upon targeted clinical research."[33]

"The problem is that we don't know enough about basic science," said Michael Brown, whose Nobel Prize with Joseph Goldstein was awarded for defining the receptor for LDL, the "bad" cholesterol.[34] Although the stimulus for research was patients with exceedingly high levels of cholesterol who sustained myocardial infarctions at an early age, their discoveries came about through basic science. "Decreasing support for basic science won't cure diseases," Brown said.[34]

"NIH should be supporting research at all levels, from the most fundamental to population-based studies," declared Eugene

Braunwald.[II] Although "NIH clinical trials initially set the bar for such research at a very high level, the attraction of performing them has decreased." Developing clinical studies has become much more complicated and has increased the administrative work of clinical investigators in both the intramural[35] as well as the extramural program. Furthermore, administrative barriers interfere with the participation of intramural investigators in multi-center trials supported by extramural funds.[36]

"We're not the NSF [National Science Foundation] since we have a commitment to clinical research, as well as basic," said Griffin Rodgers, a basic and clinical investigator who is the director of the National Institute of Diabetes and Digestive and Kidney Diseases (NIDDK) and an authority in sickle-cell disease.[37]

Also complicating the performance of human-based studies, both in the intramural and extramural programs, are the demands of the regulatory climate, the institutional review boards (IRB),* and ethical considerations, all of which make such studies, as one NIH clinical investigator said, "incredibly expensive."[38] Since investigators and their medical schools cannot afford to lose money performing these studies, they hope that the NIH will fund them more generously.

Nevertheless, partly because of the costs and difficulties involved, scientists understand why the institutes favor basic research. Much of it is less expensive, and many, particularly basic scientists, see basic studies as more innovative and valuable than most clinical studies.[38,39]

As at many medical schools, promotions and tenure at the NIH seem to come sooner to investigators working in the laboratory rather than with patients.[40] As for clinical research on drugs, "the pharmaceutical houses should do them," Braunwald, like Marks, said; "not the NIH." But, added Braunwald, there is an exception. "There are important clinical trials that sometimes need to be conducted on generic drugs that have no industrial sponsor, which the NIH should support."[II]

Zerhouni adds, "The NIH has to do drug research when industry cannot and science shows opportunity.... My view is that one should not be too dogmatic about what NIH should or should not do because no one seems to have the magic answer. You need to assess each and

---

*All institutions conducting research on human subjects must have institutional review boards (IRBs) that study proposals to make sure that they guarantee the rights and welfare of people participating in clinical trials.

every opportunity on its own merit and allow a diversity of approaches and people to try things."[1]

Roy Vagelos,[*] an academic internist who became president of the Merck Research Laboratories and chief executive officer of the company, feels even more strongly that drug trials should be conducted by the pharmaceutical houses and not by the NIH. He considers large trials to establish additional indications for drugs already approved by the Food and Drug Administration a "waste of the NIH's money. The companies would and should do them."[42†]

Other problems, according to former NCI director Vincent DeVita, include a lack of understanding of clinical research by many basic scientists, the presence of poor clinical research that negatively affects the appreciation of good clinical research, and that the predominant method of extramural funding, the R01 grant, is not well suited to most clinical investigators.[29]

The death throes of clinical investigation, lamented for decades by academic leaders,[32,43–47] prompted Harold Varmus to form "the NIH Director's Panel on Clinical Research" in 1995.[27,48–50] Careful review of a large sample of abstracts from the NIH showed that about one-third of the subjects were clinical. "No emergency here," declared David Nathan from Harvard, chairman of the panel.[51] Others agree.[52‡]

As a result of the panel's deliberations, the NIH established several awards[**] to support clinical investigators and developed collaborations with foundations to further sustain this work. Elias Zerhouni saw what the panel had done as a band aid. "He wanted a revolution."[51] Nathan fears that the Zerhouni regime is letting the one to three ratio for clinical research drop.[54] Some at the NIH say it has fallen to 15 percent.[39]

---

[*] M.D., Columbia; medical residency, Massachusetts General Hospital; NIH trainee and investigator;[41] chairman, department of biochemistry, Washington University in St. Louis.

[†] Physicians can prescribe drugs for illnesses other than those for which the Food and Drug Administration (FDA) has given its approval. The companies, however, may not publish or promote drugs for such "off label" prescribing until the FDA has approved them for specific use for particular diseases after appropriate trials and review.[42]

[‡] Varmus comments, "The idea that clinical research got a bum rap during my directorship may be held by a couple of your most emphatic interviewees, but is not widely shared."[53]

[**] Mentored Patient-Oriented Research Award (K23), Midcareer Investigator Award in Patient-Oriented Research (K24), and Clinical Research Curriculum Award (K30).

Alan Schechter, a veteran physician-scientist at the NIH, who has long lobbied for more support of clinical research, wrote, "I do not believe that this panel report is correct in its conclusions about the proportion of NIH resources devoted to clinical research or strong enough in its recommendations to reverse a real and serious imbalance."[55] Schechter considered the report a "whitewash. Grants," he said, "have been biased toward laboratory or basic science since the 1970s. We missed the chance to increase clinical research during the doubling. I believe that as many advances have come from applied as from basic research.[39] NIH stands for the National Institutes of Health, not the National Institutes of Biomedical Research, or the National Institutes of Basic Biomedical Research."[56]

"When funds are down, toxic sentiments appear," observes David Korn of Harvard University, a former medical school dean. "The war between basic versus clinical research is absolutely predictable."[57]

"Stop Blaming NIH"

Co-winners of the Nobel Prize,* J. Michael Bishop, chancellor of the University of California, San Francisco, and Harold Varmus, director of the NIH from 1993 to 1999, wrote in 2006 that the source of the angst among investigators should be directed elsewhere, not at the NIH and its Roadmap. "The NIH is a victim, not a culprit," they wrote.[59] "Shelving" the Roadmap, as advised by Marks, would not solve what are perceived to be NIH's problems, Bishop and Varmus believe. "But it just might persuade Congress and other potential critics that members of the biomedical research community are hopelessly inured to change and less concerned about the commonweal than the professional well-being of scientists. [They should] "redirect the hue and cry to Congress and the White House.... This is a time for concern and action, not despair."[59]

"We mustn't have dissention in our ranks when Congress is asking why the NIH is not producing more bang for the buck," cautions Edward Benz, chair of the Advisory Board for Clinical Research in the Clinical Center. "They're even questioning the sanctity of peer review since it includes no provision for distributing funds on a geographical

---

* In Physiology or Medicine in 1989, for research on genes that cause cancer, or, as the Nobel Committee announced, "for their discovery of the cellular origin of retroviral oncogenes."[58]

basis, obviously important for their constituencies."[60] Samuel Thier, professor of medicine and health care policy at Harvard, adds, "It wasn't helpful to undercut the NIH in this difficult time."[61]

Many extramurally funded investigators look on the NIH as "a pot of money they can draw from," in the opinion of Milton Corn, the associate director of the National Library of Medicine. "They shouldn't blame Roadmap. The problem is the shrinking budgets."[16]

CHAPTER THREE

The Institutes*

THE SIZE OF the 20 institutes varies greatly. In fiscal year 2007, the budgets ranged from $4,798 million in the National Cancer Institute (NCI) to $136 million in the National Institute of Nursing Research, and the differences in the number of employees between the larger and the smaller institutes can be substantial. Consequently, the amount and depth of the science can vary greatly among the institutes. The smaller institutes experience considerable difficulty doing "big science," so when they collaborate with a large institute like the NCI, their leaders can feel submerged by the power of the larger institute with its comprehensive infrastructure.[1]

Despite their differences in size and budgets, the institutes have much autonomy[2]—in some areas more than the director of the NIH has[3]—and, in the opinion of several past and present members of the NIH, they are quite hierarchical, each with a distinctive culture.[4] Their independence has led observers to call them "mega-silos,"[3] which form Congress tends to perpetuate.[5]

Leadership

"The vitality of the institutes depends on their directors," says Guy McKhann, former chief of neurology at Johns Hopkins and associate

---

* For more details than are discussed in this chapter about institute programs, see the appropriate section in the NIH website: http://www.nih.gov/icd/index.html.

director for clinical research at the National Institute of Neurological Disorders and Stroke in 2000 and 2001.[2] Whereas scientists in medical schools and research institutes apply for and receive funds through the extramural program, the directors receive the funds for their institutes directly from Congress and, consequently, can, with the scientific directors, exercise a great deal of control over how the money is spent in the intramural program. In this respect, "it's more like the European system," said a former institute director.[6]

The jobs appear to be quite stable. Over the years, few directors have been relieved of their positions.[7] When a recent NIH director considered changing the leadership of one of the institutes, he quickly learned that the institute's advocates and professional constituency would have prevented it.[7]

Harold Varmus, director of the NIH from 1993 to 1999, believes that institute directorships should not be considered lifetime entitlements. He told a reporter, "The most healthy situation would be for people to come and do those jobs for five or ten or twelve years. Less than five years is probably too short a time to have an imprint. As in any way of life, change is usually a good thing."[8] Accordingly, Varmus instituted five-year reviews of institute directors.[7] "I have a group of five or six people who go out and solicit opinions from a very broad swath of people who are affected by the institute, and then I and the chair of the committee get back to the institute director the opinions that have been collected," he said in an interview.[8]*

The institutes have immense convening power.[12–14] "Scientists will show up three weeks from now for a conference," said Francis Collins,

---

* The authority of institute directors is illustrated by the experience of two well-known investigators. "If one's research is not in favor with the powers that be, you don't have too many options," said Claire Fraser-Liggett, who worked at the NIH from 1985–1992. "They told Craig [Venter, then her husband], that his research had to be done on the brain since he was in NINDS [National Institute of Neurological Disorders and Stroke], and that he would have to work on brain till the end of his career. As for me, I couldn't do molecular biology in NIAAA [National Institute on Alcohol Abuse and Alcoholism] because the director wanted his institute to emphasize behavioral issues. That was a big reason that we left."[9]

For the Venters' post-NIH adventures, see *The Genome War* by James Shreeve.[10] Since 2007, Fraser-Liggett has been director of the Institute for Genome Science at the University of Maryland School of Medicine in Baltimore.[11] Venter is the founder, president, and chairman of the board of directors of the J. Craig Venter Institute, Rockville, Maryland.

Chapter Three: The Institutes   43

director of the National Human Genome Research Institute (NHGRI) from 1993 to 2008. "That's an incredible advantage. I borrow all the brain power I can. People are honored to be asked."[12] "It's hard to say 'no,' when someone is invited to an NIH conference," adds Gerald Fischbach, former director of the National Institute of Mental Health.[13]

Directors and Other Executives

The senior officials in most of the institutes are:[15]

- *Director*, who reports to the Director of the NIH. Some of the large institutes have associate directors in charge of important programs such as the HIV/AIDS program in the National Institute of Allergy and Infectious Diseases (NIAID).
- *Deputy Director*, who fills in for the director when he/she is away, and whose specific responsibilities are defined by the director and may vary greatly from institute to institute.
- *Director of Intramural Activities*, also known as the "Scientific Director," who is in charge of the intramural program at the Bethesda and other campuses of the NIH.
- *Director of Clinical Programs*, also called "Clinical Director," who supervises the mechanics of the clinical research conducted within the NIH, often at the Clinical Center, and reports to the scientific director or in some institutes to the institute director. Clinical directors, "doctors-in-chief" for their institutes,[16] supervise the clinical care that members of their institutes deliver in the Clinical Center and the training of the clinical associates. They have administrative responsibilities similar to chief medical officers in other hospitals. They also approve research protocols for studies on patients in the Clinical Center and have oversight responsibilities about NIH policies on conflicts of interest among investigators.[16] The clinical directors are members of the medical executive committee that meets twice a month with John Gallin, the Clinical Center's director, to discuss, among other subjects, hospital policies and operations, budget, credentialing, and disciplinary action.[16]*
- *Director of Extramural Activities (DEA)*, who supervises the management of grants and directs reviews for funding projects

---

* Institutes without intramural programs, like the National Institute of General Medical Science (NIGMS), do not have scientific or clinical directors.

developed within the institutes. The institutes review, on average, about 25% of applications, with the fraction varying among the different institutes. The extramural applications reviewed within the institutes, however, consume more of the extramural money than do the investigator-initiated R01 applications evaluated in the Center for Scientific Review (CSR)* because the projects developed within the institutes tend to be larger. Since the members of the institute program staffs have a vested interest in seeing that their projects are well funded, the NIH tries to maintain a "fire-wall" between them and the evaluators within the institutes.

- *Director of Extramural Programs.*[17] The institutes also have "program divisions," the number varying among the institutes. The members of the program staffs decide which grants will be funded, based on the findings of the peer-review process; evaluate the progress of the grants and contracts that the institutes have given; and develop institute-specific initiatives such as program announcements, requests for applications (RFA), and requests for proposals (RFP). The review staff organizes the peer reviews that are conducted within the institutes, and may participate in research supported by cooperative agreements and contracts. As many as 30 review groups may operate in a large institute like the National Cancer Institute. Members of the grants-management staff, who are expert in financial subjects and usually do not have doctorates, formally make the awards and arrange the transfer of funds.

Number of Institutes and Centers

NIH directors have consistently tried, usually unsuccessfully,† to prevent the proliferation of new institutes.‡ Harold Varmus, director from

---

* See Chapter 4.
† James Shannon, one of the NIH's most respected directors (1955–1968), resisted the formation of any new institutes.[18] He failed on three occasions, however: the founding of the National Institute of General Medical Science (1962), of the National Library of Medicine (1956), and of the National Institute of Child Health and Human Development (1962).
‡ For example, Phyllis Greenberger, president and CEO of the Society for Women's Health Research, and her colleagues and supporters have tried, so far unsuccessfully, to persuade the NIH to establish an institute or center that would be more effective and forceful than the current Office on Research on Women's Health (ORWH) in the NIH director's office.[19] Greenberger believes that the attention the Society brought to a better awareness of women's health contributed to the appointment of Bernadine Healy as the NIH's first and, so far, only woman NIH director (1991–1993).[19]

1993 to 1999, even tried to *decrease* the number. He proposed replacing the current structure with five categorical institutes of approximately equal size and budgets, plus NIH Central, the headquarters of the NIH, led by its director:[8,20]*

1) National Cancer Institute
2) National Institute for Brain Disorders, incorporating six current institutes such as National Institute of Mental Health and National Institute of Neurological Disorders and Stroke[21]
3) National Institute for Internal Medicine Research
4) National Institute for Human Development
5) National Institute for Microbial and Environmental Medicine
6) NIH Central

"Any attempt to eliminate individual institutes will probably meet very strong political resistance," said John Porter, the former Illinois member of Congress and avid NIH supporter.[22] An institute director added, "Reorganization of several institutes into fewer is not worth the trouble."[23] Accordingly, the 2003 study of the NIH by the Institute of Medicine hedged on recommending restructure of the institutes and centers.[24]

"I think it was obvious," Varmus comments, "that this proposal was made only for heuristic purposes, to illustrate what is wrong with the current organization. I never thought the plan would be followed for simple political reasons. But I did want to provoke debate, stop the proliferation, and have some kinds of consolidation effected."[25]

Some institute leaders agree that the organization of the NIH into multiple institutes and centers has disadvantages. "Separate institutes don't seem to make as much sense now as they may have in the past," said one director.[26]

Among those opposing consolidations of the institutes and centers are the advocates who support NIH research in particular diseases.[22] Although no action was taken on Varmus's suggestion, he can take comfort in the provision in the NIH Reform Act of 2006 that the number of institutes and centers may not exceed the current number of 27.

---

* For further details, see page 7 in "Enhancing the Vitality of the National Institutes of Health: Organizational Change to Meet New Challenges." Available at http://books.nap.edu/catalog/10779.html.

Some contend that the NIH itself is too large, that is has "bureaucratic elephantiasis ... and holds a near monopoly on finance for the biomedical sciences." Writer Daniel Greenberg advises that "a breakup of NIH into separate governmental philanthropies for the medical sciences would introduce the vigor of competition into a sector that constantly flagellates itself for scientific conservatism and operational sloth—without correcting either."[27]

Advisory Councils and Committees[17,28,29]

Each of the 20 institutes and four of the seven centers have national advisory councils composed of scientific and public members. The Center for Scientific Review (CSR), the Center for Information Technology (CIT) and the Clinical Center (CC) do not have councils because they do not fund grants.[30]

Although the composition of the councils differs slightly among the institutes and centers, their function is the same. Councils perform the second level of peer review of grant and cooperative agreement applications that have been judged fundable by primary peer review in the Center for Scientific Review or the institutes.

The councils confirm the "pay-lines," the scores needed for grants to be funded at that particular time and in that particular institute. The pay-lines vary among the institutes since priorities for funding differ among the institutes.

When the NIH was young and the number of applications many fewer than today, council members read, discussed, and evaluated each one before giving their approval for funding. In more recent decades, council members haven't the time to review each grant recommended for funding in detail; the National Cancer Institute, for example, may submit hundreds for council review. However, the council members always study the awards that the institutes want funded whose scores are below the pay-line.

In summary, each application:[31]

1. Is received in the Center for Scientific Review and assigned for review by a study section in the Center for Scientific Review or an institute or center.
2. Those receiving potentially fundable scores are reviewed by the program staffs in the institutes and centers.
3. High-scoring applications are then studied by the institute's or center's advisory council.

4. The director of the relevant institute or center then formally authorizes that the applications that are selected be funded.

Some institutes rely on the council members for advice on new initiatives and whether to spend large amounts of money on particular programs. The councils perform this work in the closed portion of their meetings because what they discuss can have proprietary interest for nonmembers.

The councils also advise the institute directors on their policies, missions, and goals. However, the institutes and centers are not obliged to follow such advice from the councils.

Part of each council meeting is open to members of the public who may hear and comment on issues relevant to the institute or center. The institute director often starts the open portion of the meeting with a review of recent institute activity, plans for the future, and budgets and other financial issues.

Most councils have 18 voting members and meet for two days, three times per year.* Although the Secretary of The Department of Health and Human Services or the President officially appoints the members for four-year terms, the institute directors and the director of the NIH nominate potential members. The chairs of the councils serve two-year terms. Membership in each council must, by law, be two-thirds scientists and one-third representatives of the public. They must include people representing different points of view and, although no quotas are given, ethnicity, gender, and geographic location are considered to give appropriate balance to the councils.[29] "Members [should] be appointed to these advisory groups because of their ability to provide scientific or public health expertise to the review and approval of awards and policies," the Institute of Medicine advised in its 2003 report on the NIH. "They should not be selected to advance political or ideological positions."[32]†

Members are appointed as "special governmental employees," are reimbursed for expenses, and given an honorarium of $200 per day. "It's less than some other HHS [Department of Health and Human

---

*Per Public Health Service Act, Sec. 406. See: "Selection criteria for NIH advisory committees," accessible at http://www1.od.nih.gov/cmo/committee/SelectionCriteria2007.pdf.
† It is the impression of one senior official that "since Bush, there have been more political appointees without adequate knowledge, but less so recently."

Services] agencies pay, but given the size of our program, it's what our budget can handle," explains Jennifer Spaeth, director of the Office of Federal Advisory Committee Policy.[29]

Whereas Congress has ordained that each of the institutes and centers that fund grants have national advisory councils,[*] the directors can decide whether their institutes and centers also have program advisory committees. As of December, 2007, a total of 32,360 individuals are members of at least one of the NIH councils or committees. Forty-four hundred are standing members, and the others serve in temporary roles on issues of special emphasis. The NIH has more advisory committees than any other department or agency in the federal government.[29]

Members of the public may apply to serve on NIH advisory committees.[33] One of these is the Council of Public Representatives (COPR). Established in 1998 by Harold Varmus, COPR, as stated in its website, "advises the NIH Director on issues affecting the broad development of NIH programs, outreach activities, and research goals. The 21-member Council is composed of members of the public from varying backgrounds, including patients, family members of patients, health care professionals, scientists, health and science communicators, and educators."[34] COPR meets twice per year in Bethesda. The members serve three-year terms. Those who fail to be selected may become COPR Associates, a group that provides "public input and feedback to the NIH."[34]

National Cancer Institute (NCI)

Established in 1937
2008 budget (in millions): $4,831[†]

Intramural program: 243 tenured senior investigators, 64 tenure-track investigators[‡]

---

[*] Through the Federal Advisory Committee Act of 1972 (FACA), a public access statute.[29]
[†] For fiscal year 2008, which began on October 1, 2007, for each institute and center.
[‡] As of October 1, 2007, for each institute and center.

Administration

In two respects the National Cancer Institute, the NIH's first institute and still its largest, is administratively "more equal" than the other institutes. President Nixon's "War on Cancer" led to the passage of the National Cancer Act of 1971, which established the National Cancer Program. Among its provisions was that the director of the National Cancer Institute would be appointed by the President,* whereas the Secretary of the Department of Health and Human Services appoints the other institute directors. The NCI director presents his budget—called the "bypass budget"—directly to the White House rather than through the NIH director as do the other institutes.[35]

Despite these apparent differences between the NCI and the other institutes, "the administrative structure doesn't really mean much," said Michael Gottesman, director of the intramural program. "They have the same procedures for hiring, promotions, and salaries, for example."[36] Richard Klausner, NCI director from 1995 to 2000, agrees. "A lot more is made about the NCI organization than applies. I thought of myself as a regular institute director. We do have a very powerful tool, outside of the budget process, to articulate opportunities and needs, since only NCI can communicate directly with Congress, which other institute directors cannot."[14]

Like the other institutes, the NCI has a council, which is called the National Cancer Advisory Board. The President appoints 18 of its members, again unlike in the other institutes, where the Secretary of DHHS appoints the members of the councils. Not more than 12 of the 23 members of the board may be scientists or physicians, and not more than eight may be representatives of the general public. All serve six-year terms. Five members of the Advisory Board are *ex officio*.[37]

The President's Cancer Panel is a small group that advises the president about the National Cancer Program.[38] In the spring of 2008, the members were: LaSalle Befall, Jr., former chairman of the department of surgery at Howard University College of Medicine; Margaret Kripke, executive vice president and chief academic officer of the University of Texas M.D. Anderson Cancer Center in Houston;

---

*The NIH director is also a presidential appointment but requires senatorial confirmation, whereas appointment of the NCI director does not.

and Lance Armstrong, the champion cyclist and cancer survivor and advocate.[39]

Along with being the largest institute or center, the NCI is, in many respects, the most complex.[40]* It is responsible for developing and conducting research in the more than 230 diseases[41] that cancer can produce in adults and children. NCI sponsors 40 percent of the research that takes place at the Clinical Center.[42] Like the other institutes,[43] NCI has its own technology transfer center.

The NCI supports research in public health through its Division of Cancer Control and Population Sciences. Robert Croyle, director of the division in the extramural program, explains, "I believe we should be more closely aligned with what the health care system needs. We're expanding into health economics, delivery of health care, particularly in local health-care settings." Croyle supervises how funds supporting this mission are spent in the extramural program. "I try to support research that captures the health of the whole person."[1]

The NCI's extramural program receives about 82 percent of the budget; the intramural, about 15 percent.[42] Supporting its work is a multitude of advocacy groups.[40]

NCI–Frederick

Unlike most institutes, the NCI operates intramural laboratories in locations other than Bethesda. The largest is at Fort Detrick in Frederick, Maryland, about 40 miles north of the main campus in Bethesda.† President Nixon established it in 1972 as part of his "War on Cancer."‡

NCI–Frederick is a federally funded research and development center (FFRDC) that is owned by the government and operated by contractors. Because of its FFRDC designation, NCI–Frederick "has greater opportunities to partner with biotech and biopharma

---

* See the NCI's comprehensive website: http://www.cancer.gov.
† There are two campuses within the gates at Fort Detrick. One is NCI–Frederick. The other is the National Interagency Biodefense Campus (NIBC) where research in support of homeland security is conducted.
‡ The NIH also operates satellite campuses at the Bayview Campus, Baltimore, Maryland (National Institute on Aging); Research Triangle Park, North Carolina (National Institute of Environmental Health Sciences); and Rocky Mountain Laboratories, Hamilton, Montana (National Institute of Allergy and Infectious Diseases).

companies," said Craig Reynolds, director of scientific operations there.* "We can do things that couldn't be easily done by grants or contracts through the extramural program or within the intramural program and can speed the development of products to diagnose and treat cancer and AIDS."⁴⁴ NCI–Frederick is the only FFRDC in the Department of Health and Human Services and the only one dedicated solely to biomedical research.†

NCI–Frederick has about 3,000 employees. Of these, 150 are principal investigators, most of whom are either tenured at the NIH or on the tenure track.⁴⁴

Richard Klausner

When President Clinton appointed Richard Klausner the eleventh director of the NCI in 1995, the new director found "a variety of real and perceived problems. NCI had a 'top-down' hierarchy that did not comport with science or medicine—a government bureaucracy with Byzantine customs."¹⁴ NCI's role in the general cancer community, Klausner felt, was unclear except as a provider of extramural funds. "Advocates found it passive, antagonistic, and without a strategic mission. I sensed that they hungered for change—did the institute have a human face?"¹⁴ The gregarious Klausner would provide it.

After training as a yellow beret in the NCI, Klausner had joined the staff of the National Institute of Child Health and Human Development as a research scientist in molecular and cell biology.‡ Explaining why he had accepted the invitation of Donna Shalala to exchange full-time work in a lab for this role in leading the largest NIH institute,** he said, "I was restless and really wanted to help the NIH institutionally." Since he had been working on cancer genes,

---

* Ph.D., microbiology, Iowa; Fellow, NCI, NIH.
† Other FFRDCs include the Argonne National Laboratory, Lawrence Livermore National Laboratory, Los Alamos National Laboratory, Brookhaven National Laboratory, Oak Ridge National Laboratory, National Defense Research Institute—each in the Department of Defense—and NASA's Jet Propulsion Laboratory.
‡ "I also moonlighted for a few years at Montgomery General Hospital," he said.¹⁴
** Since the directorate of the NCI is a presidential appointment, the Secretary of the Department of Health and Human Services nominates the candidate to the President. "Harold actually sort of stayed out as Shalala was choosing between me and Mike Bishop," Klausner explains.⁴⁵ Bishop was Varmus's co-Nobel laureate.

the appointment at the NCI seemed fitting. "I also wanted to reconnect as a doctor, and, as a research scientist, was curious to find out if planning for science was valuable or was an oxymoron."[14]

Klausner kept his lab while directing the institute. "There wasn't enough time to provide detailed supervision, and the research output fell off, but I continued to review the data. I didn't need to be an author and own the research any longer. I think it's good for directors to continue to participate in research in some useful way."[14] Accordingly, Klausner's meetings with his colleagues could be lengthy because "he got into the meat of things," said Robert Croyle.[1]

By 2000, Klausner was getting tired and restless again. "Harold [Varmus] had left, and I was close to him. At a certain point you're either an NIH 'lifer' or you're not."[14] Klausner looked at a university presidency and was under consideration to succeed Varmus. He decided instead to leave the government and a few months later joined the Bill and Melinda Gates Foundation as executive director for their global health programs. Since 2005, Klausner has been managing partner of The Column Group, a biotech venture capital group, and president of Klausner Consulting in Seattle.

Andrew von Eschenbach

The appointment of Klausner's successor was more controversial than many choices for institute directors and resulted from the special provision for the NCI in which the director of the institute is a presidential appointment. The other directors are appointed by the Secretary of the Department of Health and Human Services, who often asks the NIH director, who may have conducted a traditional academic search, for advice.

Unlike many of his predecessors, Andrew von Eschenbach had never worked or trained at the NIH when President George W. Bush appointed him its twelfth director in 2002. For 26 years, von Eschenbach,* a urological surgeon, had worked at the M. D. Anderson Cancer Center in Houston, with which the Bush family has had a long association.[46] Since he was chosen and appointed without an external

---

*M.D., Georgetown; intern in medicine, Philadelphia General Hospital (University of Pennsylvania service); surgery and urology residencies, Pennsylvania Hospital (Philadelphia); urological oncology fellow, M.D. Anderson Hospital.

search or advice from NIH officers,[47] many at the NIH saw the appointment as primarily political rather than scientific.*

The new nominee had had a distinguished career as a clinician and clinical investigator whose work was frequently supported by the extramural program of the NIH. When appointed, he had published more than 200 articles, books, and book chapters, had served as an editorial board member of several leading journals and organizational boards, and was president-elect of the American Cancer Society. At M.D. Anderson, he had held several senior clinical and administrative appointments including, from 1994 to January of 2002, the Roy M. and Phyllis Gough Huffington Distinguished Chair in Urologic Oncology.

Described as a "people-person," von Eschenbach was sociable and extroverted and devoted himself to creating "team spirit."[1] Running the institute along "business school lines,"[48] he led the NCI with the help of a small senior management group rather than Rick Klausner's practice of involving many more members in policy.[1] Accordingly, von Eschenbach was seen as leading the NCI "from 30,000 feet," and emphasizing broad strategic policies and its public face rather than operating on the ground as was Klausner's style.[1] Not a basic scientist like most of the other institute directors, von Eschenbach suffered by comparison with his predecessor, who was admittedly "a very hard act to follow."[1]

While questioning the wisdom of his appointment, NIHers hoped that von Eschenbach's close relationship with the President would help continue the NIH's generous funding. However, with the priorities affected by the September, 2001 attacks and the political preferences of the administration, he inherited the onerous responsibility of explaining to his colleagues that the days of the doubling were over.[1]

The von Eschenbach tenure was not free from controversy.[49] The Bush administration had removed from the NIH website the report of a study showing that there was no established evidence that abortion caused breast cancer. Under pressure, the NCI substituted a report that said that

---

* During the 1990s, many at the NIH had prized having an academic leader like Donna Shalala—who came to the position of Secretary of the Department of Health and Human Services from being chancellor of the University of Wisconsin–Madison— who would appoint a veteran NIH basic scientist like Varmus to be director of the NIH; and a president like Clinton who would support what she did. They denigrated, by comparison, the lack of dedication to medical research of the politicians who were to follow her as DHHS secretary and the priorities of the president who succeeded Clinton.

"abortion caus[ing] breast cancer was an open question."[50] Congressional[50] and newspaper objections[51] and an NCI-convened conference of experts[52] led the NCI to update the website to reflect the absence of an association between abortion and breast cancer.[50]

Von Eschenbach's announcement that the institute's goal was "to eliminate the suffering and death due to cancer by 2015" brought a vigorous response from scientists who thought such statements hurt more than helped research and treatment of cancer. Paul Nurse, the president of Rockefeller University, wrote, "It is no good exaggerating what science can deliver.... When we fail ... as we surely will with such a claim, we will lose the confidence and trust of both the politicians and the public."[53]

Von Eschenbach served as NCI director until June 10, 2006, when he became Commissioner of the Food and Drug Administration (FDA). He describes how his job changed: "I got a phone call from the White House asking me to come over to the FDA. President Bush wanted me to be the commissioner."[54] For six months he was both acting director of the FDA and director of the NCI,* a time of administrative unease in the institute with the director spending more of his time at the FDA than at the NIH.[1]

"I miss the NCI and a lifetime investment in cancer. I had come to the NCI because I believed we should provide leadership to change the outcome of cancer," von Eschenbach said. As for his latest job, von Eschenbach sees it as continuing his involvement in cancer. "The FDA has a critical role, since it must approve the use of cancer drugs."[54]

John Niederhuber

The next director of the National Cancer Institute was John Niederhuber,† who had been working at the NIH for a short time when he was appointed. Like his predecessor, he is an oncological surgeon. "Andy [von Eschenbach] had tried to recruit me as his deputy, but it wasn't the right time for me to move," said Niederhuber. His wife had just died from metastatic breast cancer, and their son was still in high school.[40]

---

* Von Eschenbach was the third FDA Commissioner of the George W. Bush administration. His predecessor had served slightly more than two months,[55] and a replacement was needed quickly.[47]
† M.D., Ohio State; surgical residency, Michigan.

President Bush had named Niederhuber chairman of the National Cancer Advisory Board, and when his son graduated, he agreed to spend more time at the NCI. When President Bush asked von Eschenbach to direct the Food and Drug Administration, Niederhuber became acting director of the NCI. An in-house search at the White House led to Niederhuber's being appointed director in September, 2006. One of the first issues he faced was dealing with the constricted budget that was afflicting the NCI along with the rest of the NIH.[56]

Niederhuber would like the NCI to perform more of its clinical trials at the community level of the program at nearby Suburban Hospital, where the National Heart, Lung, and Blood Institute conducts clinical research.[40*] He had to commute from Wisconsin to the NCI for an experimental drug his wife was taking. He remembers her saying, "This should be fixed so these drugs will be available at other centers and hospitals."[40]

Niederhuber favors using federally funded research and development centers for such community studies. FFRDCs are independent, private, nonprofit entities that provide scientific research and analysis for the federal government. They can perform certain projects more quickly and economically than can federal agencies like the NIH. A government agency can contract with an FFRDC to perform a particular job without having to directly commit to the personnel needed for the work. The National Cancer Institute, which manages the FFRDCs for all the institutes and centers, has found, for example, that FFRDCs can conduct research based in communities more efficiently than can the institute itself.[40]

Niederhuber also wants the NCI to conduct more follow-up studies to learn what long-term ill-effects the drugs used to treat cancer may produce.[40]

National Heart, Lung, and Blood Institute (NHLBI)

Established in 1948

2008 budget (in millions): $2,938

Intramural program: 51 tenured senior investigators,
11 tenure-track investigators

---

* See below.

The National Heart, Lung, and Blood Institute, one of four institutes established in 1948, was first called the National Heart Institute. It became the National Heart and Lung Institute in 1969 and received its current name in 1976.* Marshall Nirenberg, chief of the Laboratory of Biochemical Genetics in the intramural program, won the Nobel Prize in Physiology or Medicine in 1968 for deciphering the genetic code. Nirenberg was the first NIH and the first federal employee to receive a Nobel Prize.†

Elizabeth (Betsy) Nabel,‡ an academic cardiologist, well known for her work on gene therapy for coronary artery disease and formerly chief of the cardiology division at the University of Michigan, has been described as very businesslike, very political, and very smart. She came to the NIH, as she acknowledges, as the "spousal recruit."[58] The NIH was trying to persuade her husband, Gary Nabel, a molecular virologist and immunologist known for his research in HIV, cancer, and the Ebola virus, to become director of the vaccine research center in the National Institute of Allergy and Infectious Diseases. To complete the deal, the NIH offered Elizabeth Nabel the new position of scientific director of clinical research in the National Heart, Lung, and Blood Institute. She accepted, the Nabels moved to Bethesda in 1999, and six years later she succeeded Claude Lenfant as director.**

Clinical Programs

When Nabel arrived, the intramural program in clinical cardiology was moribund. The halcyon days of Eugene Braunwald in medical

---

* Claude Lenfant, a pulmonary physician who was director of NHLBI for 21 years (1982–2003), remembers when the lung societies tried to create a separate institute for their specialty. "Ted Cooper [institute director from 1968–1974] prevented it."[57] Cooper argued successfully that the heart and lung were intimately related, both anatomically and functionally, and that research in both fields at the NIH should be jointly administered.[57]

† The research accomplishments of the NHLBI are many and can be reviewed on the institute's website: http://www.nih.gov/about/almanac/organization/NHLBI.htm.

‡ M.D., Cornell; resident and fellow, Brigham and Women's Hospital.

** Consistent with her work in coronary artery disease is Elizabeth Nabel's interest in the rare Hutchinson-Gilford progeria syndrome, the main feature of which is premature aging.[59] Patients with this condition die at a mean age of 13 from myocardial infarction or stroke due to accelerated, premature atherosclerotic disease caused by loss of the smooth muscle cells in arteries. Francis Collins and associates in the National Center for Human Genome Research at the NIH helped determine the genetic basis of the disease.[60]

cardiology and Glenn Morrow (1922–1982)[61-63] in cardiac surgery in the 1960s were long gone. Braunwald had been succeeded by his former fellow Stephen Epstein, who directed the cardiology branch for 30 years, but in 1998 he had departed for the MedStar Research Institute in Washington, DC.

In 2001, Toren Finkel,* a cardiologist who had "become more a molecular investigator than a clinician," as he explains, was appointed to succeed Epstein. On October 1, 2007, Finkel's title changed to chief of the newly created translational medicine branch, in keeping with director Zerhouni's emphasis on this subject, and to reflect Finkel's research. His responsibilities now include the former cardiology and pulmonary branches and the laboratory of cardiac energetics. Finkel's group is studying the therapeutic benefits of stem cells in cardiovascular disease.[64] As for the clinical work of the cardiology branch, "the NHLBI will try to find someone, but we anticipate there may be the expected problems with recruitment," Finkel said.[64]

Cardiac surgery, both clinically and in clinical research, had never flourished as much at the NIH as when Glenn Morrow was the chief.[61,62] Morrow was deeply committed to the NIH† and willing to accept less income than he could have earned at a university or in private practice to work at the NIH. Cardiac surgeons are among the most highly paid doctors, and recruiting and retaining them at the NIH with its limitations on salary and other restrictions became very difficult. Only a surgeon dedicated to a research mission,[61] like Morrow then and Stephen Rosenberg in the cancer institute now, or one with a private income would consider taking the job.

The absence of an emergency department at the Clinical Center discouraged patients with heart disease from coming to

---

* M.D. and Ph.D., Harvard; resident, Massachusetts General Hospital; cardiology fellow, Johns Hopkins.

† "One of the reasons Glenn remained at the NIH," remembers cardiac surgeon Charles McIntosh, "was the pleasure he derived from training four or five new clinical associates, who would spend two years in a program divided between caring for cardiac surgical patients and participating in cardiovascular research. Glenn trained over 130 clinical associates during his tenure, many of whom became department or division chairs in academic centers all over the U.S."[65] McIntosh was a clinical associate from 1968 to 1970 and worked at the NIH until he retired in 1990.

the NIH. Furthermore, other institute directors are less than enthusiastic about supporting an expensive cardiac surgery program in the Clinical Center.[58] Under the "school tax" system that former NIH director Harold Varmus and John Gallin, the director of the Clinical Center, established, each institute paid a portion of its budget to support the Clinical Center regardless of how much its investigators used it.

Suburban Hospital

Recognizing the impossibility of resurrecting a cardiac surgery program in its historical form at the NIH—no cardiac surgery had been performed there since 1990—Nabel undertook a unique approach. One block behind the NIH campus on Old Georgetown Road is Suburban Hospital, a not-for-profit community hospital founded in 1943. It operates 238 beds, has a daily census of 170 to 190 patients[66] and about 15,000 admissions each year. Of the 900 doctors who have privileges, about 350 admit most of the patients.[66] Since the hospital imposes no limit on what they can earn, a cardiac surgeon could make a competitive salary working there.

Nabel, Lenfant, colleagues in the NHLBI, and Zerhouni envisioned a collaboration between the NIH and Suburban in which cardiologists and cardiac surgeons with academic interests would perform their clinical work at Suburban and conduct research at the NIH.

Shepherding the project from concept to operating took seven years, despite the help of Johns Hopkins.[67,68] Obtaining a certificate of need (CON) was the most frustrating step. The State of Maryland requires that a hospital acquire a CON to build the facilities and operate invasive cardiology and cardiac surgery programs. Accordingly, the state announced an open competition for another hospital in the region to build a cardiac surgery program. In addition to Suburban, whose interest had stimulated the competition, three other hospitals applied. In December, 2002, the Maryland Health Care Commission unanimously selected Suburban,[69] but this was only the first step in what became a prolonged and at times bitter fight.

Adventist Hospital, another community hospital with a cardiac surgery program, sued the Maryland commission to reverse the decision. MedStar Health, a large not-for-profit corporation that owns the Washington Hospital Center, the largest hospital in the District of

Columbia, and the Georgetown University Hospital* joined the suit to protect its large cardiac surgery program at the Hospital Center. The suit reached Maryland's highest court, which rejected the commission's decision on a four-to-three vote.[71] Suburban revised its application, reapplied, and again the commission awarded it a CON. MedStar and Adventist appealed a second time, but this time the court sided with Suburban, also by a four-to-three vote.[72]

On September 29, 2006, the NIH Heart Center at Suburban Hospital opened as "a place to study normally sick people,"[66,72] in contrast to the unusual patients who are studied in the Clinical Center at the NIH. The facility had cost Suburban $16 million.[73] In its first year, 223 coronary artery bypass graft operations and cardiac valve procedures were performed. The NIH referred 10 percent of the patients.[74]

Each of the cardiologists and cardiac surgeons whom Suburban and the NIH recruited through national searches is committed to running a clinical program with significant translational research. Much of the population in the region tends to be highly educated and high earners, so, as Eugene Passamani,† who worked at the NIH from 1972 to 1993, said, "our population is not afraid of research."[66]

The director of cardiothoracic surgery is Keith Horvath,‡ who came from the Northwestern University School of Medicine in Chicago.[58] Horvath's salary is paid from a practice plan developed by Suburban. This arrangement solves the NHLBI's chronic problem of not being able to develop an academic cardiac surgery program because of the limitation on salaries under which the NIH operates.

Because Horvath arrived before Suburban's program opened its surgical suites, he operated at the Johns Hopkins Hospital for nine months.[67] He had even accepted the job before the CON controversy had been finally settled. "Keith took a huge chance coming here when he did," said Suburban chief executive officer Brian Gragnolati.[72]

---

* See Chapter 9 in the author's *Selling Hospitals and Practice Plans: George Washington and Georgetown Universities*.[70]

† M.D., Michigan; intern, Massachusetts General Hospital; resident and fellow in cardiology, Barnes Hospital, St. Louis. Passamani started talking with NHLBI about a joint cardiology program at Suburban as early as 1996, and is credited by many with first suggesting the joint program.[72] He is now senior vice president for research and education at the hospital.

‡ M.D., Chicago; residency, Brigham and Women's Hospital.

Horvath performs animal research in a laboratory at the NHLBI where colleagues conduct the research program under Horvath's direction.[74]* Horvath works at the NIH under the provisions of the Intergovernmental Personnel Act (IPA) which permits non-NIH employees to participate, without compensation by the government, in activities that support the mission of the NIH. Since Horvath is not an NIH employee, he has no administrative responsibilities there.[58] The NIH reimburses Suburban for the time Horvath spends there as an investigator.

Since 1999, the NIH and Suburban have been associated in another program to study heart disease with sophisticated scanning techniques. The NIH has installed expensive magnetic resonance imaging (MRI) equipment in space provided by Suburban, which has also assigned offices for those working on the project.

"Suburban allows us to work with patients with garden-variety heart disease who have been admitted through its emergency department to the coronary care unit and cardiac catheterization labs," said Andrew Arai, the NIH senior investigator who directs the program.† "Without an emergency department, we seldom see such patients at the NIH. What we do have is a team of engineers and technicians who are studying how to make the images more practical and useful. The patients benefit from the studies as we learn how to best apply the new techniques."[75] Arai, who spends much of his time performing research in laboratories at the NIH, can study the findings that his colleagues collect on the patients across Old Georgetown Road through electronic linking between Suburban and the NIH. "It's a unique environment," said Arai, "and the work has such promise."[75]‡

As for the arrangement with Suburban, Eugene Braunwald said, "What Betsy engineered is nothing short of brilliant."[76]**

---

* Horvath's research projects include angiogenesis, MRI-guided cardiac surgery, and xeno-transplantation of hearts from pigs into baboons.[74]
† M.D., Illinois in Chicago; house officer, Oregon: cardiac fellow, Oregon and NHLBI.
‡ NHLBI is not the only institute with a program at Suburban Hospital. NINDS (National Institute of Neurological Disorders and Stroke) conducts a stroke study there.[43]
** In April, 2009 Johns Hopkins Health System announced that it would acquire Suburban Hospital.[76a]

> **National Institute of Allergy and Infectious Diseases (NIAID)**
>
> Established in 1948
> 2008 budget (in millions): $4,583
>
> Intramural program: 98 tenured senior investigators, 24 tenure-track investigators

In 1981, Anthony Fauci was chief of the laboratory of immunoregulation in the National Institute of Allergy and Infectious Diseases. He had come to the NIH as a yellow beret in 1970 after graduating from Cornell University Medical College and training in medicine at the New York Hospital. Except for one year as chief resident at New York Hospital in 1971 and 1972, Fauci would become an NIH "lifer," and in 2008 at age 64, he is still there. Appointed director in 1984, he is by far the longest-serving institute director. Certified in three disciplines—internal medicine, infectious diseases, and allergy and immunology—and having conducted research in the regulation of the human immune system for a decade, Fauci was especially prepared to study the cause of an unexplained acquired immune deficiency disease that was appearing predominately in gay men. "It was a stroke of fate that I was trained to be an AIDS doc before AIDS came along," he said.[77]

AIDS Comes to the NIH and NIAID

The disease, initially called GRID for "gay related immune deficiency," was first reported in June of 1981 in a group of homosexual men in New York City and California who presented with pneumocystis pneumonia, an unusual lung infection, and with Kaposi's sarcoma, a similarly uncommon cancer.[78] When it was recognized that the immune system of patients with the disease was severely malfunctioning, Fauci realized that his experience could be applied to the new entity. When on service at the Clinical Center as an infectious diseases consultant, he had seen a very ill patient with immunodeficiency, which, in retrospect, must have been an example of the new syndrome. By the fall of 1981, Fauci decided to recruit patients with the syndrome to the NIH and start a program to study them.

At the beginning, the cause was unknown, but to Fauci, the infectious disease doctor, as well as to other specialists,[79] GRID looked like a viral infection. When heterosexual men using intravenous drugs also appeared with the disease, the case for an infection was strengthened. Fauci first reported his concept of the disease early in 1982.[80]

The first patients Fauci and his colleagues studied were "deathly ill" with what was now being called AIDS, "acquired immunodeficiency syndrome." "It became a full-time thing to take care of them," he remembered. "They were so ill that admission to the intensive care unit was essential."[77] There Henry Masur,[81] an infectious disease fellow, whom Fauci had known when Masur* was a medical student at Cornell, became what Fauci called "our AIDS guy in the ICU."[77] Philip Pizzo, a pediatrician in the Cancer Center, who would later become chief of pediatrics in the National Cancer Institute (NCI) and of medicine at Boston Children's Hospital and then dean at Stanford, cared for the infected children.†

Robert Gallo, then working in the NCI, suggested that AIDS could be caused by a class of viruses known as retroviruses.[79] He and his colleagues developed the technology to find the human immunodeficiency virus (HIV), and with co-discoverer Luc Montagnier from the Institut Pasteur in Paris showed that HIV caused AIDS.[84] Gallo's group also developed a test to diagnose more precisely who had the disease and thereby prevent its spread though transfusions of blood products.

Working with a pharmaceutical house, Samuel Broder, also at the NCI and later its director (1989–1995), found that the drug AZT (azidothymidine), long on the shelf after proving to be an unsatisfactory cancer drug, could, at least temporarily, reduce the replication of the virus.

Pizzo wanted to try using the drug for children with AIDS. "We had to leap over many hurdles to be able to try AZT clinically," Pizzo remembers. "The drug company wasn't initially interested in

---

* Masur is now director of critical care medicine at the NIH.[81]
† One NIH employee, a technologist in the blood bank, cut her finger handling blood from an AIDS patient. Eighteen days later she had a flu-like syndrome from the virus. She eventually developed HIV/AIDS and died eight years after being infected. "As you'd expect, the staff became very worried," remembers David Henderson, the hospital epidemiologist, and also the deputy director for clinical care in the Clinical Center now. "I walked from unit to unit telling them what we had learned, and we quickly instituted appropriate preventive measures."[82,83]

using it for pediatric studies, and we also had to convince the FDA (Food and Drug Administration) to let us use it." Pizzo remembers that Broder was studying AZT, not as a cancer drug, but because, as an immunologist, he wanted to learn more about its effects on retroviruses, of which HIV is one, as Gallo had predicted. "This is a good example of how lines can productively blur between disciplines at the NIH and how the intramural program can mobilize rapidly to enter a new area of research. Much tougher to do this in the extramural world."[85]

The early difficulties defining many of the scientific features of AIDS and the lack of effective treatment led its advocates, a particularly vocal and effective group, to claim that "nothing was being done," remembers William Paul, who was to become a senior NIH administrator for research in HIV/AIDS.[86] In 1993, Congress, responding to the advocates' pressure, established, against the wishes of the NIH leaders, the Office of AIDS Research in the Office of the Director of NIH.

Harold Varmus tried to persuade Bernard Fields, a leading virologist and then a department chairman at Harvard, to direct the office, but Fields had developed cancer and had to withdraw. Varmus then chose Paul, an immunologist, but, like Fields, not a specialist in HIV/AIDS, to be director of the office and associate NIH director for AIDS research. With advocates insisting on "better coordination" of AIDS research, Paul was given the authority to bring together AIDS research in all the institutes, a distinct administrative departure from the convention that each institute director had total authority over the research in his institute.[86] Paul directed the program for four years; "after that I went back to the lab."[86] The annual budget was about $1.3 billion. "We put a lot of money into creating a vaccine. Still don't have one," he said."[86]

Fauci, meanwhile, concentrated on defining the immune deficiency and providing clinical care for the adult patients. Becoming recognized as a leading and articulate authority on the subject, he advised the NIH leadership and each of the subsequent presidents and senior members of their administrations about HIV/AIDS, the nomenclature Fauci prefers.[77] "A master entrepreneur and politician who can express the science in terms that Congress and the laity can understand," is how one of his colleagues describes Fauci. "Tony tells it like it is and is very careful to adhere to the data."[87]

Among his many honors, Fauci became the second NIH intramural scientist to win a Lasker Award, the most respected American honor in

medical science, in 2007 for his work in HIV/AIDS and biodefense.[88]*
In 2008, President George W. Bush presented the Presidential Medal of Freedom to Fauci.

Vaccine Research Center

The ultimate dream of potential patients and research workers in HIV/AIDS is the development of a vaccine to prevent becoming infected. Despite intense study of the disease for almost 30 years, an effective vaccine has not yet been developed.[89] One technical problem has been that the virus "is a moving target,"[90] constantly genetically mutating so that the vaccines developed so far soon lose their potency. The practical problem was the absence of a single laboratory where enough trained and highly qualified investigators, supported with enough resources, could be brought together to work on the problem.[90]

The NIH's role in this quest began when William Paul, the NIH administrator for research in HIV/AIDS, suggested that an intramural vaccine program emphasizing HIV/AIDS be established at the NIH. "I mentioned this to Clinton," said Harold Varmus, "and he included the proposal in his commencement talk at Morgan State [University, Baltimore, in 1997.] We then got funds appropriated quickly and set a record for construction of a federal facility."[25]

In April, 1999, after a search that Varmus, who ran the search, calls "not an easy one,"[25] the NIH recruited Dr. Gary Nabel,† then at the University of Michigan, to direct the Vaccine Research Center. In 2000, the Center moved into its newly constructed building, specifically designed to facilitate the research. "We can change scientific directions quickly, much easier than in the extramural program," said Nabel, a former recipient of extramural NIH grants, "and the infrastructure is wonderful here. We network a lot with extramural scientists, the CDC [Centers for Disease Control and Prevention], the

---

* The first Lasker winner in the intramural program was Harvey Alter from the Clinical Center (2000) for discovering the virus that causes hepatitis C and developing screening methods to reduce the risk of hepatitis transmitted by blood transfusion.

† B.A., M.D., Ph.D., Harvard; residency in medicine, Brigham and Women's Hospital; fellow, Dana-Farber Cancer Institute and laboratory of David Baltimore, Whitehead Institute, Massachusetts Institute of Technology. Baltimore is the chairman of the center's advisory group.

military, the Gates Foundation, and industry, providing much cross-fertilization."⁹⁰

Administratively, NIH assigned the Center to the National Institute of Allergy and Infectious Diseases, thereby bringing Fauci's scientific and administrative experience to assist in its development. By 2008, the Center included nearly 200 internal federal or contractor employees at the NIH—nine senior investigators and one tenure-track investigator—and an additional 150 external, contract employees.*

The budget for the 2008 fiscal year is $101.6 million. "The flat budgets have prevented us from developing some programs," Nabel said, "but we haven't had to close any labs."⁹⁰ Unfortunately, despite the knowledge gained at the Vaccine Research Center and other centers at universities and in industry, Nabel believes that another ten years may be required before a vaccine will be available that will consistently prevent infection of humans by HIV.⁹⁰ Robert Gallo compared the failure of a major HIV/AIDS vaccine trial to "the Challenger space shuttle disaster." The reporter for *Science* magazine wrote that Fauci said at a meeting he convened, "NIAID needs to set a new course for a field that seems to have hit a brick wall."⁹²

National Institute of Dental and Craniofacial Research (NIDCR)

Established in 1948
2008 budget (in millions): $392

Intramural program: 15 tenured senior investigators, one tenure-track investigator

---

The National Institute of Dental and Craniofacial Research is the only government agency that funds significant amounts of oral research. "We deal with disparities among dentistry across all countries," said Lawrence Tabak, the director.⁹³ The institute he leads and the National Institute of

---

* Myron Levine, director of the Center for Vaccine Development at the University of Maryland School of Medicine, calls Nabel's unit "a great operation in a building with superb laboratory facilities infrastructure. Gary Nabel has recruited outstanding scientists to staff the Vaccine Research Center."⁹¹

Nursing Research (NINR) are the only institutes at the NIH assigned to specific professions.

"Roughly 20 percent of the population suffers from roughly 80 percent of the disease burden of caries," said Tabak. "Did you know that dental decay is mankind's most common infection?"[93]

Given that most oral cancers are discovered in the dentist's office, the dental institute, even more than the National Cancer Institute, deals with the consequences of head and neck cancers.[93] For example, radiation treatment of these lesions can produce salivary dysfunction, so some institute scientists concentrate on saliva technology. Others study pain in the head and neck. They are trying to better understand and develop treatments for temporomandibular joint dysfunction, a painful condition of the joints between the lower jaw (the mandible) and the temporal bone of the face that the institute has found to occur in five to fifteen percent of people. Facial development, including such conditions as cleft palate, is also an interest of investigators in the dental institute.

National Institute of Diabetes and Digestive and Kidney Diseases (NIDDK)

Established in 1950
2008 budget (in millions): $1,716

Intramural program: 81 tenured senior investigators, 18 tenure-track investigators

---

The history of the National Institute of Diabetes and Digestive and Kidney Diseases provides a good example of how several of the institutes have become amalgamations of different medical diseases and organs.

First called the National Institute of Arthritis and Metabolic Diseases (NIAMD) in 1950, the new institute incorporated the laboratories of the Experimental Biology and Medicine Institute (founded in 1948) and expanded to include research in rheumatic diseases, diabetes, and several metabolic, endocrine, and gastrointestinal diseases. In 1972, NIAMD became the National Institute of Arthritis, Metabolism, and Digestive Diseases, and in 1981, the National Institute of Arthritis, Diabetes, and Digestive and Kidney Diseases. In 1986, the institute's activities in arthritis, musculoskeletal, and skin diseases were spun off to become the core of the new National Institute of Arthritis and Musculoskeletal and Skin Diseases (NIAMS). What was left became the current NIDDK, which is responsible for research into such varied conditions as diabetes and nutrition, and diseases of the gastrointestinal system, kidneys, urological system, endocrine organs, and hematological systems.

Advocates continue to press for the creation of separate institutes for urology, nephrology (kidney diseases), GI (gastrointestinal tract), and diabetes. "It's more difficult now thanks to the recent reauthorization," said Harold Varmus.[25]

Diabetes

Among the valuable studies that NIDDK sponsored is the clinical trial testing whether strict control of blood sugar in patients with type 1 diabetes would decrease the incidence of small-blood-vessel (microvascular) complications that can lead to blindness, kidney failure, and neurological disease. The data, announced in 1993, clearly showed that patients who maintain tight control of blood sugar have fewer of these complications.[94] NIDDK has also helped support a study in the United Kingdom that showed that microvascular complications are also reduced in the much more common type 2 diabetes in patients who adhered to a strict regimen. Long-term follow-up of both study populations showed that the microvascular benefits continued long after the period of improved blood sugar control ended, and cardiovascular disease benefits also emerged over time.* This is the kind of study at which the extramural program excels. Pharmaceutical companies tend not to be interested in funding studies that require many years of observation or involve the use of more than one drug.[94]

To guide patients in controlling the disease, NIDDK joined with the CDC to create the National Diabetes Education program (NDEP) in which more than 200 public and private organizations participate.[†]

---

* Type 1 diabetes is caused by loss of the insulin-producing beta cells in the pancreas. Insulin is the hormone that controls the metabolism of blood sugar. Since type 1 diabetes often affects young patients, it was formerly known as "childhood diabetes." Type 2 is much more common and develops in patients of all ages but usually older than those with type 1 diabetes. Type 2 is principally caused by resistance of the body's cells to the effects of insulin.
† See http://ndep.nih.gov/ for more information about NDEP

Patients with diabetes are advised to maintain comprehensive control of blood sugar, blood pressure, and cholesterol to levels proved by clinical trials to reduce their risk of diabetes complications. The NDEP expanded its efforts to the estimated 57 million Americans who have "pre-diabetes" after another study, supported with extramural funds from the NIDDK, showed that favorable changes in activities of daily life delay the appearance of diabetes in those with a tendency to develop the disease.[94]*

Polycystic Kidney Disease

One of the most important studies now being conducted with NIDDK sponsorship is research in genetic polycystic kidney disease (PKD).[31]† The cysts replace nephrons, the structures in the kidney that filter the blood and produce urine. When the kidneys fail to function adequately, the patient develops end-stage renal disease (ESRD) and must have dialysis.‡ About three percent of patients on dialysis have PKD. Healthy kidneys transplanted into these patients do not develop polycystic disease.

NIDDK is sponsoring work that may lead to a diagnosis earlier in the lives of affected patients so that treatment can be applied earlier. Some patients enter the study because they have abdominal pain that imaging shows to be due to polycystic kidney disease. Others enter because they have relatives whose kidney disease has been diagnosed. If one of the parents has the most common form of PKD, on average one of every two children in the family will develop the disease.

Traditionally, doctors have used blood tests of kidney function to alert them that kidney disease is present. However, since PKD develops gradually, the diagnosis may not be made until much of the kidney function has been destroyed. In NIH-sponsored studies now under way, investigators are measuring the size of kidneys with PKD using magnetic resonance imaging (MRI). One study found that the cysts enlarge the kidney in an orderly and largely predictable fashion before kidney function deteriorates; hence, measuring the size of the kidneys

---

* See http://www.diabetes.org/pre-diabetes.jsp for more information about pre-diabetes.
† See http://kidney.niddk.nih.gov/kudiseases/pubs/polycystic/ for more information about polycystic kidney disease and the NIH.
‡ "Renal" is adapted from the Latin word for kidney.

may enable physicians to follow the progression of PKD early in the course of the disease. The NIH studies are now determining if increase in kidney size predicts later deterioration of kidney function; if so, measurement of kidney volume might dramatically lower the size, length, and cost of future studies of treatment.[95]

Other studies may reveal whether drugs that protect the kidneys in hypertension—which PKD often produces—will prevent or at least slow the progression to end-stage renal disease.

National Institute of Mental Health (NIMH)

Established in 1949
2008 budget (in millions): $1,413

Intramural program: 42 tenured senior investigators,
13 tenure-track investigators

The National Institute of Mental Health (NIMH) has had a complicated organizational history since it was founded on April 15, 1949.[*] In 1967, NIMH was separated from the NIH and given bureau status within the Public Health Service. A year later it became a component of the Health Services and Mental Health Administration (HSMHA) also within PHS. The NIMH intramural research program, however, remained at the NIH. In 1970, the National Institute of Alcohol Abuse and Alcoholism (NIAAA), and in 1972, the National Institute on Drug Abuse (NIDA) were established within NIMH. After briefly rejoining NIH in 1973, NIMH was reassigned to a new agency, the Alcohol, Drug Abuse, and Mental Health Administration (ADAMHA) which succeeded HSMHA.

In 1992, Congress assigned all research components of NIAAA, NIDA, and NIMH into the NIH as three separate institutes, while the service, as distinguished from the research, components of each became part of a new PHS agency, the Substance Abuse and Mental

---

[*] For the early history of NIMH, see Gerald N. Grob, The National Institute of Mental Health and mental health policy, 1949–1965, in C. Hannaway, ed., *Biomedicine in the twentieth century: Practices, policies, and politics*. Washington, DC: IOS Press, 2008, pp. 59–94.

Health Services Administration (SAMHSA), and that is the structure that currently exists.[96]*

Herbert Pardes

Herbert Pardes[†] was chairman of the department of psychiatry at the University of Colorado when he was selected to become the fourth director of the NIMH in 1978. He remembers that of the 130 candidates for the position, committees pared the list down to 10, then to two. The finalists were then interviewed by Hale Champion, the Under-Secretary, and Joseph Califano, the Secretary of the Department of Health, Education, and Welfare, and other government leaders. Pardes was offered and accepted the job with plans to change the policies of the institute. He was buoyed by the support of First Lady Rosalynn Carter, who had a strong interest in mental health.[97]

"I wanted the institute to concentrate more on psychiatric illness rather than so strongly on social and global issues," Pardes said. He favored greater emphasis on evidence-based clinical care, research, and epidemiology. "The director had the authority to move the money around and reassign people, and I did. I hoped we would become like the more respected cancer and heart institutes."[97] Pardes believes that the changes he instituted helped strengthen the extramural research programs in mental health, which complemented the already-strong intramural program. Thomas Insel, the current director, remembers Pardes as "very involved in our research—very hands-on, a terrific inspiration."[98]

Pardes found that enthusiasm for supporting mental health decreased when the Reagan administration took office in 1981. "Reagan took us out of clinical work; he didn't think government should be providing such services. It was a very tumultuous time." In 1984, Pardes left the NIMH.[‡]

---

* For further information about the National Institute of Mental Health, see http://www.nih.gov/about/almanac/organization/NIMH.htm.
† M.D., State University of New York, Brooklyn; intern and resident in psychiatry, Kings County Hospital, Brooklyn; training in psychoanalysis, New York Psychoanalytic Institute.
‡ Pardes then became chairman of the department of psychiatry at Columbia University College of Physicians and Surgeons. He was subsequently named dean of the medical school and is now president and CEO of New York-Presbyterian Hospital.

Chapter Three: The Institutes   71

Steven Hyman

The eighth director of the NIMH was Steven Hyman,[*] whom Harold Varmus appointed from Harvard in 1996. "Exactly the right guy for the time," said Alan Leshner, who had co-chaired the search committee. "He blew us away during the search process."[99]

Hyman, a psychiatrist, ran a laboratory at the Massachusetts General Hospital where he studied the molecular biology of neurotransmitter action. Hyman's commitment to basic science appealed to the NIH director, who wanted a young scientist working in neurobiology to direct the institute.[99]

"It was a heady time to start," Hyman remembers. "Despite his shutting down the government, Gingrich was enthusiastic about science, and the NIH stayed open. The doubling [of funding] was coming, and we had Varmus and many talented colleagues to work with. Donna Shalala [Secretary of Health and Human Services] appreciated research, and we had the strong support of John Porter and Arlen Specter in the Congress. The president [Bill Clinton], who sensed that science matters, was benign toward the NIH and left us alone."[100]

Hyman moved from directing a research laboratory and Harvard's Mind, Brain, and Behavior Initiative into administering an institute with an annual budget of $900 million. "Quite a change," he observed. "I was an unusual choice, 43 years old, and, though trained in clinical psychiatry, I'd never published in a psychiatric journal."[100]

Hyman thought that much of what the NIMH was doing at the time was not relevant to understanding the basis of psychiatric illness.

> I wanted to involve the NIH in more genetics research, since most psychiatric illnesses have a genetic basis. We needed more work on the fundamentals of neural functioning. There were very few data about the biology of mental illness in children, and the drug company trials were too short for the chronic illnesses characteristic of our field. I decreased the emphasis on social determinants of psychiatric illness.[100]

---

[*] M.A., Cambridge University (history and philosophy of science); M.D., Harvard. Intern in medicine, Massachusetts General Hospital (MGH); resident in psychiatry, McLean Hospital; clinical fellow in neurology, MGH; postdoctoral fellow in molecular biology, Harvard.

This change in emphasis was not received with equanimity at academic centers where more conventional, nonbiological, psychiatric research still dominated. Not everyone supported Hyman's belief that the NIH should sponsor large-scale clinical trials of drugs for treating psychiatric illness. "There's an important role for the federal government in being an impartial, unbiased investigator of certain sensitive areas," he told a reporter from *The New York Times*.[101] These trials were needed, he believed, to address questions the drug industry neglected or had a vested interest in.[101] Hyman was also criticized for encouraging advocacy groups that were questioning the ethical soundness of studies on psychiatric patients. "Good ethics and good research are inseparable," he said.[101]

In 2001, Hyman left the NIH. "I couldn't abide the growing politicization of things. Science has to be free. Science and politics don't mix."[100] Hyman returned to Harvard as provost of the university.

Thomas Insel

Thomas Insel,[*] whom Elias Zerhouni recruited as the ninth director the year after Hyman left, was committed to "continuing Steve's trajectory. We've 'discovered' the brain, so we've become a neuroscience institute."[102] Insel remembers how the institute before Hyman emphasized service, social psychology, and psychotherapy.

> ...without much of a tie to neuroscience. We're a disease-specific institute now. We want to understand the biology and pathophysiology of mental illness. Unlike in the rest of medicine, many of our diseases occur in patients who are outside the medical system. Large collections of those with serious mental illnesses are in the criminal justice system, on the streets as homeless people, and in nursing homes where many are depressed and inadequately treated. Fifty per cent of the deaths from nicotine addition, I suspect, can be ascribed at least partly to mental illness.[102]

---

[*] M.D., Boston University; psychiatric residency, Langley Porter Neuropsychiatric Institute, University of California–San Francisco; clinical associate, National Institute of Mental Health.

Insel and his colleagues are frustrated by how little is known about the mechanisms of the major diseases that NIMH deals with. "We have therapies that help most patients get better but not well, and we lack the knowledge about many psychiatric disorders to develop preventions and cures," he says.[102]

NIMH has increased the portion of its budget assigned to the extramural program. It now sponsors large-scale trials to measure the effectiveness of drugs over relatively long periods of time, which are required to evaluate drugs used for mental illness. "We are studying chronic illnesses, so our trials must be longer than the efficacy studies conducted by drug companies," says Insel. "For example, about 30 percent of patients with depressive disorders respond to the first medicine they receive, but it takes eight to twelve weeks before we can really know whether the therapy helped, and we know too little about long-term remission."[102]

The NIMH exercises more control over the grants it funds than do many other institutes. Even applications with good scores may not be funded if their subjects fall outside the institute's strategic plan.[102]

Autism

NIMH has become the lead institute for studies of autism, research on which is conducted in four institutes in addition to NIMH. "There's a new public focus on autism," said Insel. "It's become like AIDS or juvenile diabetes, with large amounts of private money and intense public lobbying for public research dollars."[98]*

The focus in autism research was formerly on behavioral research, explains Insel. "Now we think of autism and related illnesses as developmental brain disorders. The behavioral symptoms may be a late manifestation. Unfortunately, we only make the diagnosis behaviorally at a relatively late stage. This is analogous to using a heart attack to diagnose coronary artery disease."[98]

Like several other institutes, the National Institute of Neurological Disorders and Stroke has had several names and functions since it began, as the National Institute of Neurology and Blindness in

---

* Though autism advocates have lobbied for an independent institute, the NIH Reform Act of 2006 prohibits the creation of any more institutes and centers. The Act, however, did establish an Interagency Autism Committee.[29]

> **National Institute of Neurological Disorders and Stroke (NINDS)**
>
> Established in 1950
> 2008 budget (in millions): $1,552
>
> Intramural program: 37 tenured senior investigators, 12 tenure-track investigators

1950.[103] In 1968, the blindness program was extracted to become the nucleus of the National Eye Institute, and what remained was renamed the National Institute of Neurological Diseases and Stroke. In 1975, it became the National Institute of Neurological and Communicative Disorders and Stroke (NINCDS). Finally, in 1988, the present name was adopted when the communicative disorders program became the focus of the National Institute of Deafness and Other Communication Disorders (NIDCD).

As stated on its website, "The mission of NINDS is to reduce the burden of neurological disease—a burden born by every age group, every segment of society, and people all over the world."[104] The institute's work has been facilitated with the recent opening of the Porter Neuroscience Research Center,[105] named in honor of former congressional representative John Porter, one of the NIH's most enthusiastic supporters. Institutes that spend at least 50 percent of their budgets on neuroscience research—such as the institutes that study neurological disease, mental health, eye diseases, deafness, aging, and drug abuse— have laboratories there. "Since the NIH does not have a neuroscience institute, we have, in effect, created one, emphasizing collaborative work, in the Porter building," said Story Landis, director of the neurological institute. "We agreed to create a pool of $100 million over five years to support our extramural work as well. Elias [Zerhouni, the NIH director] calls it a blueprint for the brain."[106]

Zach Hall

Harold Varmus recruited Zach Hall* to be the sixth director of NINDS in 1994 from the University of California–San Francisco (UCSF),

---

* Ph.D., Harvard; fellow in biochemistry, Stanford.

where Varmus was working before becoming director and Hall was chairman of the department of physiology.

"I told Harold that I wanted to do something in the public service," he said.[6] When Varmus and Hall worked at UCSF, they ran together each Saturday afternoon "to the ocean,"[6] and continued this tradition in Bethesda when Hall was not commuting to his home in San Francisco, where his wife plays the English horn in the San Francisco Symphony. Hall and Varmus brought the informal dress style of California to the NIH.

Hall, a neuroscientist and the first director of the institute with a Ph. D. and not a medical degree, found his institute in trouble. "The program officers weren't contemporary scientists, so the extramural program lacked up-to-date scientific relevance." Advancement to tenure had become routine for most with sufficient time in service. "It was an inbred promotion system," Hall said. When he arrived, about 20 promotions to tenure were being considered. "I stopped most of them, advertised for the positions and instituted searches. I was told that I wouldn't be able to make any changes, that the NIH was a bureaucracy like a pillow absorbing a punch."[6]

Despite the NIH's being more hierarchical than universities, Hall found it easier to institute decisions since one didn't have to get the approval of the faculty where collegiality reigned, about which Hall was all too familiar from his days as a department chairman at UCSF.[6] As for the threat of scientists' leaving because an administrative decision went against them, Hall said, "they couldn't pick up their grants and take them to a university."[6] Intramural investigators have no grants.

By 1997, Hall had become sufficiently weary of his long commute that he resigned his directorship and returned to UCSF as vice-chancellor for research.[6*]

Gerald Fischbach

"I was uncertain what I would find when I got there," remembers Gerald Fischbach[†] as he started a three-year term as director of the

---

[*] Hall also became a leader in the California stem cell initiative.[25]
[†] B.A., Colgate; M.D., Cornell; intern, University of Washington, Seattle; fellow, National Institute of Child Health and Human Development (NICHD).

NINDS in 1998, "but I had confidence in Harold Varmus. There's great joy in doing scientific administration at a place like the NIH, where you know you're performing a worthwhile service."[13] Fischbach agreed with Varmus's preference that institute directors run research laboratories in addition to administering their institutes.*

Like other institute directors, Fischbach found that he could make the changes he thought wise and didn't have to constantly answer to the NIH director for his decisions. One of his first moves was to abolish the three divisions he inherited: stroke, neurodegenerative diseases, and clinical trials. He created clusters of subjects, not named after diseases: cognition and behavior, synapses and circuits, non-neural cells (tumors are part of this group) and neurogenetics, and "I gave more power to the clinical people."[13]

During the Fischbach administration, he decreased the amount of money allocated to intramural research from 12 percent to 10 percent, thereby assigning slightly more to investigators at institutions outside the NIH. Most of the extramural funds were allotted in keeping with the priority scores developed by the study sections, but the institute exercised discretion in assigning about 20 percent of the money to develop particular areas of research.[13] Fischbach strove to develop relationships between investigators in the extramural and intramural programs and at other institutes.

In 2001, Fischbach left the NIH after only two and a half years. He concluded that "an era was ending. Varmus was leaving, and a new administration had been elected."[13]†

---

* Varmus had appointed Fischbach, then chairman of the department of neurobiology at Harvard, director of NINDS under the Intergovernmental Personnel Act (IPA), whereby he remained on the Harvard faculty, was paid by Harvard with his Harvard salary, and the NIH reimbursed Harvard. After two years, this arrangement had to be discontinued, and Fischbach became a government employee at two-thirds of his university salary.[107]

† On leaving the NIH, Fischbach became dean of the Columbia University College of Physicians and Surgeons (P&S) in New York. He is now John E. Borne Professor of Medical and Surgical Research at P&S and scientific director of the Autism Initiative, Simons Foundation, New York, N.Y.

Story Landis

The eighth director of the National Institute of Neurological Disorders and Stroke is Story Landis,* its first female director and second director with a Ph.D. rather than an M.D. degree. The NIH conducted three searches to fill the position after Gerald Fischbach left. It lost its leading candidate in the first search to a pharmaceutical house, and the second search's candidate didn't want to leave a productive research career for an administration job. When the third search started, Landis, then the institute's scientific director of the intramural program, applied and was selected, even though, she said, "the neurology community wanted a neurologist for the job."[106] Their point of view has mellowed, according to Lewis ("Bud") Rowland, former chairman of the department of neurology at the Columbia University College of Physicians and Surgeons. "She comes to all the clinical meetings and relates well to the clinical neurologists."[108]

Landis had been chairman of the department of neurosciences at Case Western Reserve School of Medicine in Cleveland when she became scientific director of the NINDS in 1995. Her research focuses on understanding how functional connections form in the developing nervous system.

Landis started as director of the institute in September, 2003, just as the funding doubling was ending and the new conflict-of-interest regulations were being developed for intramural investigators. "The first meant that it was difficult to launch new initiatives," said Landis, "while the second made it difficult to recruit physician-investigators to the intramural program."[109]

She also found herself having to balance the scientific opportunities in neurological research with the advocates' desires for cures. "Public expectations are often not consistent with the stately pace of transitional and clinical research," she said. "The public and Congress expected that the doubling of the budget would double the number of cures."[106]

During Landis's terms as scientific and institute director, NINDS recruited scientists into independent sections or units and closed several

---

*Ph.D., Harvard; postdoctoral training, National Institute of Mental Health and Harvard Medical School. "Story" is her great-grandmother's maiden name. "My first name is Elizabeth, but my mother didn't like any of the nicknames I was accumulating, so she called me Story, and it stuck."[106]

laboratories and branches while most of the institutes retained the lab-branch administrative structure. "We were most aggressive in weeding out our less successful investigators and hiring outstanding people from the outside," she said.[106]

National Library of Medicine (NLM)

Established in 1956

2008 budget (in millions): $322

Intramural program: 10 tenured senior investigators, three tenure-track investigators

---

The National Library of Medicine[110] is the world's largest medical library.[111] The collections consist of more than nine million books, journals, technical reports, manuscripts, microfilms, photographs, and images. Almost 6,000 medical libraries are connected to the NLM. The NLM is not a circulating library, but anyone, not just doctors, investigators in medical science, and medical historians, can work there.[112]

The library is housed in a striking building on the Bethesda campus of the NIH, which is "considered a prime example of Cold War architecture," according to Elizabeth Fee, chief of the History of Medicine division. The books are all stored underground, so that if the roof caves in during a attack, the books will be preserved."[112]*

Within the library is one of the world's finest medical history collections of old and rare medical books. The History of Medicine division has published on its website scientific and personal biographical sketches about leading medical scientists and lay people who have contributed to the development of medicine.[†]

Research and development in the NLM is carried out in the Lister Hill National Center for Biomedical Communications (LHNCBC) and the National Center for Biotechnology Information (NCBI), Donald Lindberg,[‡] the director, explains.[111] The LHNCBC explores the uses of

---

* The rare books were sent to Cleveland during World War II.[112]
† See http://profiles.nlm.nih.gov for the NLM's Profiles in Science site.
‡ M.D., Columbia; resident in anatomical and clinical pathology, Columbia-Presbyterian Medical Center, New York.

computer, communication, and audiovisual technologies to improve the organization, dissemination, and utilization of biomedical information.[113]

The NLM supports grants through its extramural program for research in medical informatics, health information science, and biotechnology information, as well as for research training in these areas.[113]

On the website are the following data about the NLM as of September 20, 2006:[114]

- Staff (full-time equivalents)—655
- Articles indexed (all databases)—623,000
- Circulation requests—467,000
- Collection (book and non-book items)—9,186,000
- Journals indexed (for MEDLINE)—4,900
- Serial titles received—20,800
- Titles cataloged—32,000

Internet

The library makes available its collections of medical literature over the internet, one of its most valuable services. As the NLM's website states: "For 125 years, the Library published the Index Medicus®, a monthly subject/author guide to articles in 4,000 journals. This information, and much more, is available in the database MEDLINE, the major component of PubMed, freely accessible through the World Wide Web."[113] These electronic services are coordinated through the NLM's National Center for Biotechnology Information, the major research and development arm of the National Library of Medicine for information about molecular biology. NCBI, which has about 500 employees (2008),[115] creates public databases, conducts research in computational molecular biology, develops software tools for analyzing genome data, and disseminates biomedical information.*

PubMed, a free search engine for accessing the MEDLINE database of citations and abstracts of biomedical research articles, includes links to full-text articles at several thousand journal web sites as well as to most of the articles in PubMed Central.[113] Full-text manuscripts do not appear in PubMed, but they are available through PubMed Central (PMC), which offers free access to its database of full-text journal articles.

---

* See http://www.ncbi.nlm.nih.gov/ for more information about NCBI.

PubMed does not include citations for certain types of publications, such as book reviews, that are considered beyond of the scope of PubMed. These items constitute a small portion of the total PubMed Central collection. As of 2008, the NLM had no plans to include them in PubMed.

PubMed Central contains over 1.5 million articles from more than 450 journals, many of which have a corresponding entry in PubMed.[116,117]

MedlinePlus[115] is a site that is "intended," according to its website, "to be used by health care providers and patients, and designed to provide up-to-date, authoritative information. MedlinePlus is updated daily."*

The National Center for Biotechnology Information of the NLM is also the home of GenBank®, the NIH genetic sequence database and the repository of all DNA sequences from publicly funded research. The content of GenBank doubles every 12 months.[117]

Public Access Policy[118]

The public-access policy of the NIH prescribes that the author or publisher of every scientific paper written by an NIH-funded investigator must be submitted to PubMed Central upon acceptance. PubMed Central will make it publicly available within 12 months.[117,119] The rationale for the policy is that, since the National Institutes of Health is a public agency supported by federal money, taxpayers have the right to see all knowledge developed by anyone receiving NIH support.[117,119]†

This policy is not unique to the NIH or the United States. Similar rules apply to some agencies in Britain and Canada, and to the members of the European Research Council. The Howard Hughes Medical Institute in the United States has a similar rule.[119]

"We're the Place of Last Resort"

Despite the wide use of the internet to connect to the NLM's resources, the library continues to maintain a comprehensive collection

---

* See http://medlineplus.gov/ for more information about MedlinePlus.
† See http://publicaccess.nih.gov/ for more information about the NIH Public Access Policy.

of bound journals and books. As Donald Lindberg said, "Print is wonderful and will persist. We're the place of last resort."[111]

> **Eunice Kennedy Shriver National Institute of Child Health and Human Development (NICHD)**
>
> Established in 1962
> 2008 budget (in millions): $1,251
>
> Intramural program: 83 tenured senior investigators, 15 tenure-track investigators

The National Institute of Child Health and Human Development is the principal focus for research in pediatrics and obstetrics at the NIH.* Investigators in NICHD study human development from before conception through adulthood, with the aim of improving the health of children, adults, families, communities, and populations.

The institute has "always been engaged with public issues such as whether vaccines cause autism, stem cells, and *in vitro* fertilization,"[121] said Arthur Levine, who worked at the NICHD for 16 years.† "Despite this very broad mandate, NICHD had not had a large appropriation. Nevertheless, the institute has always devoted resources and been responsive to contentious public issues such as whether vaccines cause autism and whether embryonic stem cell research is ethically appropriate. Budget cuts are a particular challenge."[122]

NICHD conducts about one third of the intramural and extramural research affecting children and women. The rest is performed in other institutes; for example, cancer in women and children in the National Cancer Institute. Among the subjects studied within NICHD, in addition to human growth and development, are prematurity, eclampsia, assisted reproductive technology (such as *in vitro* fertilization), infertility, disabilities, sexually transmitted diseases, contraception, mental retardation, genetics, pregnancy, and perinatology.‡

---

* The NIH has resisted creating a separate institute for obstetrics despite the advocacy of supporters.[120]
† Levine is now senior vice-chancellor for health sciences and dean at the University of Pittsburgh School of Medicine.
‡ See the NICHD website for further information: http://www.nichd.nih.gov/.

"We're not a disease-oriented institute," said the director, Duane Alexander. "The subjects sponsored by NICHD are among the most complex that scientists at the NIH study. Much adult disease is related to pre-birth issues."[120]

National Children's Study

The National Children's Study examines the effects of environmental exposure on child health and development.[123] The study defines "environment" broadly to include natural and man-made environmental, biological, and psychosocial factors.*

The National Children's Study is a multi-agency U.S. government project involving, in addition to the National Institute of Child Health and Human Development, the National Institute of Environmental Health Sciences (NIEHS), the Centers for Disease Control and Prevention, the U.S. Environmental Protection Agency (EPA), and the U.S. Department of Education. Congress appropriated $69 million in fiscal year 2007 and $110 million in 2008 specifically for this study.[124]

The children's exposures and their health and development will be evaluated during pregnancy—ideally from early in the first trimester—at birth, six months, twelve months, and three years, and then about every three years until the age of 21. Eventually 100,000 children will participate. The consortium will release the first results in 2012 and afterwards as they become available.[125]

Some questions that the study may answer include:

- Can very early exposure to some allergens help children remain asthma-free?
- How do genes and the environment interact to promote or prevent violent behavior in teenagers?
- Are lack of exercise and poor diet the only reasons why many children are overweight?
- Do infections affect developmental progress, asthma, obesity, and heart disease?
- How do city and neighborhood planning and construction encourage or discourage injuries or obesity?

---

* See http://www.nationalchildrensstudy.gov for additional details about the study.

Chapter Three: The Institutes 83

> **National Institute of General Medical Sciences (NIGMS)**
>
> Established in 1962
> 2008 budget (in millions): $1,946
>
> Intramural program: None

Jeremy Berg was in his thirteenth year as director* of the department of biophysics and biophysical chemistry at the Johns Hopkins University School of Medicine[125] when he took a phone call from a colleague on the search committee encouraging him to become a candidate for Director of the National Institute of General Medical Sciences at the NIH. "I had essentially used up the start-up funds I was given when I became chairman. It was time to think about doing something different," he remembers. "The more homework I did, the more intrigued I became. I felt I could make more of a difference there than by remaining at Hopkins and could give back something to the institute that had funded much of my research."[126] Berg had worked with NIH director Elias Zerhouni at Hopkins, knew him well, and admired him. Financial considerations did not matter, because, as he was a basic, not a clinical, scientist, "it was essentially a wash."[126]

So, in 2003, Jeremy Berg became the sixth permanent director of the institute about which many in academic medicine and the public know the least.† Basic medical scientists like Berg and investigators in biology, chemistry, and other scientific disciplines related to medical problems, however, know the institute well, since this is where many apply for support.

NIGMS received 6.8 percent of the NIH budget in fiscal 2006. Only three other institutes received more.[127] Almost all of NIGMS's money is spent outside the NIH. It is the only institute without a substantial intramural program. Ruth Kirschstein, Berg's predecessor,

---

* Berg (Ph.D., Harvard; postdoctoral studies, Hopkins) was a young 33 when he became a Hopkins department chairman (called "director" at Hopkins).
† Berg and his family did not move from their home in suburban Baltimore. His wife is a practicing radiologist there, and his children attend local schools. Berg leaves his home at about 6:00 A.M. to commute to Bethesda. The trip takes about an hour in the absence of traffic, which state of affairs Berg describes as " a hypothetical concept."[126]

who directed NIGMS from 1974 to 1993,* said, "I was very happy about not having an intramural program that would only compete with institutes and their programs. Why would you then want to be the director? No one dies of general medical science."[129]

Berg's personal research is conducted in the National Institute of Diabetes and Digestive and Kidney Disease. "If the lab were in my institute," he explains, "I would report to a scientific director who would report to me. Not an ideal arrangement. Besides, there are great investigators in the NIDDK to collaborate with."[126]

Function and Purpose

"NIGMS was started 45 years ago, and is one of the cleverer things that Congress did," Berg explains. The institute he leads is the only one devoted to the study of basic mechanisms of illness and not to specific diseases. NIGMS does, however, support some clinical subjects, essentially those without homes in the disease-specific institutes, such as anesthesia, burns, trauma, and wound healing. The institute has a special mission to improve training for minorities at the NIH and elsewhere. It supports research and training on problems of interest to two or more institutes, which most of the other institutes do not.

The major programs for which NIGMS is responsible include: cell biology and biophysics; genetics and developmental biology; pharmacology, physiology, and biological chemistry; bioinformatics and computational biology; and minority opportunities in research.

NIGMS has supported at some time in their careers 64 scientists who have won the Nobel Prize in Physiology or Medicine and in Chemistry, more than at all the other institutes together. The reason is that most Nobel prizes are given to basic scientists, and these are the investigators whom the institute has been designed to support. "We win all these Nobels," Berg has said as often as possible, "because of taxpayer supported medical research."[126]

"Not having an intramural program simplifies things," Berg has learned. "Our mission is clear. We don't have to decide how to allocate

---

*Ruth Kirschstein was the first woman to direct an institute at the NIH. She and her husband, Alan Rabson, who served as deputy director of the National Cancer Institute, were able to work in the same institution because the NIH did not apply the anti-nepotism rules then followed in many universities. See Buhm Soon Park's "The anti-nepotism rules in academia," in *Perspectives in Biology and Medicine* 2003;46:396–399.[128]

money between intramural and extramural programs. Our structure relieves me of running an intramural program and dealing with needy scientists and their personal ambitions," issues that took much of his time while a medical school department chairman. However, he spends more time administering in his NIH job than he did at Hopkins.[126]

Funding Grants

As in the other institutes, the leaders of the NIGMS influence how money is spent in the extramural program. Most of the grants are funded solely on the basis of their priority scores, but Berg and his associates will favor some whose scores are less competitive. "We like young investigators who are just beginning their careers," Berg said, and the institute may not approve a highly scored application for continued funding by a senior investigator whose laboratory has a large amount of other support.[126]

"I'm struck by the competence of the career administrators here," says Berg. "They take a broader view of science than most investigators running their own labs." The 45 program directors, who meet with Berg at least every month, follow the activity in their scientific fields closely. They discuss the grants being considered and develop the targeted programs the institute supports. All hold Ph.D.s and many have engaged in research earlier in their careers. Berg has concluded that "they have a lot of influence, more than I found scientific administrators exercised at the university. They spend a lot of time in meetings, but here they can really have an impact."[126]

Medical Scientist Training Program

The National Institute of General Medical Science administers, as an extramural project, the Medical Scientist Training Program (MSTP), which supports the education of medical students who want to obtain the Ph.D. degree in addition to the M.D.[130,131]* The program, which was prompted by the shortage of physician-scientists in American

---

*See http://www.nigms.nih.gov/Training/InstPredoc/Predocoverview-MSTP.htm on the NIH website for further information about the program. The director of the MSTP program at the NIH is Bert Shapiro.

medicine, started in 1964 with three institutions. Currently, there are 45 participating degree-granting institutions training about 900 students each year. About 170 positions are available each year for students to enter the program.

Students in the MSTP program pay no tuition and receive a stipend and modest support for travel, equipment, and supplies. Since the NIH does not pay all these expenses—about 60 percent of the tuitions and 90 percent of the other costs—the medical schools must contribute financially to the support of each student. The candidates are not required to pay back the money invested by the NIH and their medical schools in developing their careers as physician-investigators.

The medical schools, not the NIH, recruit and select the students. Many schools appoint and pay the costs of more M.D./Ph.D. candidates than the NIH has agreed to support. About 75 medical schools have M.D./Ph.D. programs, 30 more than have NIH support through the MSTP program. The leaders of those without MSTP programs hope that eventually the NIH will agree to include their schools among those with MSTP programs.

National Institute of Environmental Health Sciences (NIEHS)

Established in 1969
2008 budget (in millions): $646

Intramural program: 55 tenured senior investigators, 14 tenure-track investigators

---

In 2004, the NIH was looking for a new director of the Institute of Environmental Health Sciences, the only institute whose entire operation is not on the Bethesda campus. Since its founding in 1969 as a unit in the National Cancer Institute studying the environmental effects of cancer,[47] NIEHS has been in Research Triangle Park, North Carolina, where it was the first biomedical organization to be sited there.[132]* The investigator

---

* Brought about by successful lobbying by members of the faculty and officials at the University of North Carolina in Chapel Hill, the governor of North Carolina, and several members of Congress. For more details, see A.N. Link, *A generosity of spirit: The early history of the Research Triangle Park Foundation of North Carolina*, 1995; p. 88.

selected was David Schwartz,* then a professor of medicine, genetics, and environmental sciences and policy at the nearby Duke University medical school.

"I was happy at Duke when several members of the search committee and others encouraged me to apply," he remembers. Schwartz filled out the application form, emphasizing his qualifications in program development, areas of research, and record of developing the careers of trainees and junior faculty. He found the process that followed to be "very professional and not confrontational."

Ten to 15 of the 50 to 60 who had applied were invited to Bethesda. Each candidate was interviewed by a committee of 15 people prominent in the field, mostly from outside the NIH. "For one hour, I was peppered with questions. Then some of us were asked to write papers describing where we thought the field of environmental science was going and where it should go."[133]

As a finalist, Schwartz was brought back to the NIH for meetings with Director Zerhouni and other leaders at the NIH. "The whole process had been based on science, not administration. I was never asked political questions such as my opinion on stem-cell research. By the end of the process, there was little they didn't know about me. It was the best selection process I'd ever been through."[133]

Zerhouni then told Schwartz that the NIH were "very interested," but it took time for Schwartz's needs in space, equipment, lab support, moving expenses,† and compensation to be sorted out. He would be paid 60 percent of what he was earning at Duke, his university salary there having been augmented by honoraria, consulting, and other fees. These would be forbidden for him to accept once he became a government employee.‡ He also gave up the tuition benefits Duke offered for his children, who were about to enter college. "The cut was

---

* M.D., University of California, San Diego; postdoctoral training in tropical medicine and pulmonology at Walter Reade Institute of Research, Boston City Hospital, and the University of Washington; Robert Wood Johnson Clinical Scholars Program, University of Washington; M.P.H. in occupational medicine, Harvard School of Public Health.
† Not much of a problem in Schwartz's case, since the Institute of Environmental Health Sciences is only eight miles from Duke.
‡ According to *The Washington Post*, the limitations imposed by these restrictions caused Schwartz to delay accepting the position.[134,135]

worth it," he decided, and, on May 23, 2005, David Schwartz become the fourth director of the NIEHS and third director of the National Toxicology Program (NTP).

"I've got lots of people to help me," Schwartz said in explaining why he was able to balance the administrative demands of the institute with his own research interests. "As for grants, 60 percent are extramural, where we can make our greatest impact on the field by helping to decide which are funded." In recent years, NIEHS has supported between 400 and 500 R01 grants—120 new ones each year. "Our biggest job," said Schwartz, "is figuring out where to go with new initiatives for research."[133]

During his tenure at the NIEHS, Schwartz "helped create the Genes, Environment, and Health Initiative, the Epigenomics Roadmap Initiative, a translational research program supported by NIEHS, and a more comprehensive approach to training and career development."[136]

Schwartz Leaves

Schwartz's tenure at NIEHS was short, however; his appointment as director ending in February, 2008. He had been criticized for, among other reasons, receiving fees as an expert witness for asbestos lawsuits and for operating a large personal laboratory that contained guest researchers from Duke.[137,138] Although institute directors are encouraged to continue their research, administration of the institute is their primary responsibility.

Problems at the institute had caught the attention of Senator Charles ("Chuck") Grassley (R-Iowa), ranking member of the Senate Committee on Finance. In the summer of 2007, he had begun "an inquiry into allegations of mismanagement and ethical lapses at the National Institute of Environmental Health Sciences."[139] The senator's action prompted the NIH to conduct a review into deficiencies in the management of the institute and to recommend solutions.[140]

On leaving the NIH, Schwartz became director of the pulmonary division and genetics center at the National Jewish Medical and Research Center in Denver. Although declining to answer questions about the reasons for this change, Schwartz observed that, in his judgment, "Congress plays too strong a role influencing scientific decisions at the NIH."[141]

Chapter Three: The Institutes 89

> **National Eye Institute (NEI)**
>
> Established in 1968
> 2008 budget (in millions): $671
>
> Intramural program: 23 tenured senior investigators, four tenure-track investigators

Ophthalmologic research at the NIH was first assigned to the National Institute of Neurological Diseases and Blindness (NINDB). With both lay and professional advocates pressing for separate recognition at the NIH for the study of diseases of the eye, the National Eye Institute was created in 1968.[142] NINDB would eventually become the National Institute of Neurological Disorders and Stroke.

Carl Kupfer, the founding director of the Eye Institute, served for thirty years, longer than any other director in the contemporary history of the NIH. In 2001, Paul Sieving succeeded him as the institute's second director.* Sieving sees patients in the Clinical Center with the genetic form of retinal disease known as retinitis pigmentosa and Stargardt macular degeneration and directs a laboratory studying pharmacological approaches to retarding degeneration in transgenic and naturally occurring animal models.

In its extramural program, the institute supports research in six scientific areas: retinal diseases; corneal diseases; lenses and cataracts; glaucoma and optic neuropathies; strabismus, amblyopia, and visual processing; and low vision and blindness rehabilitation. There are six laboratories in the intramural program: immunology, molecular and developmental biology, retinal cell and molecular biology, sensorimotor research, neurobiology-neurodegeneration and repair, and ophthalmic genetics and visual function.

---

* M.D. and Ph.D. in engineering, Illinois; resident, University of Illinois Eye and Ear Infirmary, Chicago; fellow, University of California, San Francisco, and Harvard Medical School.

> **National Institute on Alcohol Abuse and Alcoholism (NIAAA)**
>
> Established 1970
> 2008 budget (in millions): $439
>
> Intramural program: 14 tenured senior investigators, nine tenure-track investigators

The National Institute on Alcohol Abuse and Alcoholism has had a complicated bureaucratic history. When founded as part of the NIH in 1970, NIAAA was charged "to develop and conduct comprehensive health, education, training, research, and planning programs for the prevention and treatment of alcohol abuse and alcoholism."[143] Four years later, the institute was transferred, along with the National Institute of Mental Health and the National Institute on Drug Abuse, into the Alcohol, Drug Abuse, and Mental Health Administration, an entity with bureau status within the Public Health Service. In 1992, the NIAAA came back to the NIH as a new research institute. Its clinical responsibilities during the ADAMHA days had been assigned elsewhere.

In 2002, Ruth Kirschstein, as acting director of the NIH,* recruited Ting-Kai Li,† from the University of Indiana, where he was professor of biochemistry, distinguished professor of medicine, and director of the Indiana Alcohol Research Center, as the sixth director of the National Institute on Alcohol Abuse and Alcoholism.

The intramural program receives 11 percent of the National Institute on Alcohol Abuse and Alcoholism's budget.[145] The Clinical Center admits an average of three patients a week and has an average daily census of 10.4 patients with alcohol-related diseases. Among the subjects being studied in the intramural program and supported in the extramural program are the diagnosis, prevention, and treatment of

---

* Kirschstein, who started at the NIH in 1956, was twice Acting NIH Director in 1993 and from 2001 to 2002. Kirschstein has been described as "the consummate protector of the NIH."[144]

† M.D., Harvard; resident in medicine, Peter Bent Brigham Hospital; fellow, Nobel Research Institute and Karolinska Institute, Stockholm, Sweden.

Chapter Three: The Institutes 91

alcohol addiction, and the genetics, neuroscience, epidemiology, health risks, and benefits of alcohol consumption.[143]
The institute also:[143]

- Collaborates with international, national, state, and local institutions, organizations, agencies, and programs engaged in alcohol-related work.
- Translates and disseminates research findings to health-care providers, researchers, policymakers, and the public to increase the understanding of normal and abnormal biological functions and behavior relating to alcohol use.

National Institute on Drug Abuse (NIDA)

Established in 1974
2008 budget (in millions): $1,006

Intramural program: 22 tenured senior investigators, four tenure-track investigators

---

Though the National Institute on Drug Abuse was established in 1974, it only came to Bethesda in 1992, having previously been based at what is now the Johns Hopkins Bayview Medical Center in Baltimore, among other places.

In 1994, Harold Varmus made his first selection of an institute director in Alan Leshner to head NIDA. From 1988 to 1994, Leshner* had been deputy director and for two years acting director of the National Institute of Mental Health. Leshner's goal in developing his institute was to recruit outstanding scientists so that NIDA would become the leading center for the study of the neuroscientific basis of addiction.[99] By 2001, when Leshner left, NIDA would be supporting 85 percent of the world's research on drug addiction.[99]

Given the nature of the work that NIDA performs, its director becomes involved in political controversies. As an example, Leshner wanted the ban on federal funding of needle-exchange programs lifted.

---

* Ph.D., Rutgers, in physiological psychology.

"We had concluded that the program met all the requirements for our support. It would help reduce the spread of infectious diseases like HIV/AIDS and wouldn't increase the use of drugs. We convinced Donna [Shalala], but Clinton wouldn't lift it because of politics."[99]

Leshner left NIDA in 2001 to become chief executive officer of the American Association for the Advancement of Science. Since 2003, the director has been Nora Volkow, a psychiatrist.* Before coming to the NIH she had spent several years at the Brookhaven National Laboratory in Upton, New York, and at the State University of New York–Stony Brook School of Medicine, where Volkow became professor of psychiatry and associate dean of the medical school.

Volkow uses contemporary scanning techniques to study psychiatric illnesses and, in particular, the mechanisms of addiction—for which, she said, funding for diagnosis, treatment, and clinical research are inadequate. "Much of the public think it's something other than a disease," she explains. "Most medical schools teach little about it. The pharmaceutical companies are not working to develop drugs because there's no reimbursement, the patients can't pay, and addiction has few advocates."[23]†

National Institute on Aging (NIA)

Established in 1974
2008 budget (in millions): $1,053

Intramural program: 35 tenured senior investigators,
19 tenure-track investigators

The National Institute on Aging (NIA) was established in 1974 after an intensive lobbying campaign by several interested organizations and despite the opposition of President Nixon‡ and many members of

---

* M.D., National University of Mexico UNAM, Mexico City; resident, New York University.
† See NIAAA's website for details about the research conducted in its intramural and extramural programs. http://www.nih.gov/about/almanac/organization/NIAAA.htm.
‡ Nixon vetoed a bill establishing the institute the first time it crossed his desk in 1972 but relented and signed another just before resigning as president in 1974.[146]

Congress. "The institute would have had trouble getting enough financial support if the advocates had gone away or been frozen out," said Robert Butler, the director from 1976 to 1982.[146]

The institute is charged with studying the basic science and the clinical aspects of the aging process. Most of the intramural research is conducted at the Gerontology Research Center and the Biomedical Research Center on the Johns Hopkins Bayview Medical Center campus in Baltimore. Some programs are located on the main NIH campus in Bethesda or in the nearby Gateway Building. Research in aging is also conducted in other institutes, but the NIA is usually the "lead institute" that coordinates research on particular topics related to aging.[147]

Butler remembers how, in the institute's early days, "we had to put Alzheimer's on the map. We had to convince people that about half of the 'senile' people in the back wards of nursing homes were really Alzheimer patients. It's much more common than was suspected in those days."[146]

In 1993, Bernadine Healy (NIH director, 1991–1993) persuaded Richard Hodes,* an immunologist at the National Cancer Institute, to become the third permanent director of the NIA. Hodes was a different type of director for the institute. "He's a Varmus person, not primarily a clinician or policy person," said Butler. "So there's more scientific research now and less advocacy for geriatrics [than] when Frank Williams [director, 1983–1991] and I were running around the country talking up geriatrics and convincing people to apply for grants."[146]

National Institute of Arthritis and Musculoskeletal and Skin Diseases (NIAMS)

Established in 1986

2008 budget (in millions): $511

Intramural program: 14 tenured senior investigators, four tenure-track investigators

---

* M.D., Harvard; medical intern and resident, Massachusetts General Hospital; research fellow, NCI and Karolinska Institute, Stockholm, Sweden.

The National Institute of Arthritis, Musculoskeletal, and Skin Diseases was established "from the rib of what became NIDDK [National Institute of Diabetes and Digestive and Kidney Disease]," according to Stephen Katz, the director of NIAMS since 1995. When appointed to his current position, Katz was the head of the dermatology branch in the National Cancer Institute.

Katz is one of about half of the institute directors who continue to run research laboratories, in his case studying dermatological diseases with an immunological basis. "We have the most committed, brightest, most collegial, most institutionally sensitive people you can imagine," Katz modestly describes his colleagues.[148] Throughout his career at the NIH, ten of the fellows who have worked in his lab have become chairs of departments of dermatology in the United States— one is a dean—and 18 others have assumed similar positions at international institutions.[148]

In the intramural program, NIAMS currently has about 20 investigative groups, four of which are being led by tenure-track scientists, and about 110 fellows and students. The institute receives about 1,200 investigator-initiated grants annually for funding in the extramural program, of which 20 to 25 percent have been funded per year. NIAMS also funds three to eight percent of applications with scores beyond the pay-line* that the institute believes offer unusual opportunities for creative research or are high-risk, high potential payoff research. Rarely, "three times in my 13 years," said Katz, "we reject grants that the study sections liked. In some cases, we found the applications unethical, redundant to work we were already funding, or missing an essential feature such as inadequate biostatistical support."[148]

National Institute of Nursing Research (NINR)

Established in 1986

2008 budget (in millions): $138

Intramural program: one tenured senior investigator, one tenure-track investigator

---

*The funding of applications that are beyond the pay-line is called "select pay" in some institutes.[149]

The National Institute of Nursing Research, which began as a center in 1986, acquired its higher administrative status as an institute in 1993. Nursing joined the NIH after powerful advocacy and lobbying and over the objections of the NIH leadership at the time.

NINR's strategic plan emphasizes:

- Promoting health and preventing disease;
- Improving quality of life through self-management, symptom management, and caregiving;
- Eliminating health disparities;
- Taking the lead in end-of-life research.

The institute's second director, whom Varmus appointed in 1995, is Patricia Grady.[*] "Ninety percent of our funding is clinical," said Grady. "Nursing research is relatively new, so our intramural program is small, and most of the money is in the extramural program, much of which is directed to training."[150] The institute supports several programs to prepare nurses to pursue careers in research and sponsors mentored experiences for candidates to study issues of particular importance to minorities.

National Institute on Deafness and Other Communication Disorders (NIDCD)

Established in 1988
2008 budget (in millions): $396

Intramural program: 15 tenured senior investigators, one tenure-track investigator

---

Like several of the institutes, the National Institute on Deafness and Other Communication Disorders owes its origin to the efforts of advocates, but in this case many were otorhinolaryngologists—usually called "ENTs" for "ear, nose and throat" doctors—in addition to the usual interested members of the public.[151]

---

[*] B.S.N., Georgetown; Ph.D. in physiology, Maryland. Before coming to the NIH in 1988, Grady was a faculty member of the school of nursing at the University of Maryland and of the neurology department of the school of medicine at Maryland, where she studied arterial stenosis and cerebral ischemia.

Before NIDCD was created in 1988, much of the research in this area had been conducted in what would become the National Institute of Neurological Disorders and Stroke, and the doctors felt that their specialty wasn't adequately represented in the NIH. Senator Tom Harkin, whose half-brother's hearing was impaired,[152] provided particularly enthusiastic support for creating an institute that would specialize in deafness.

The current director of NIDCD is James Battey,* whose appointment by Harold Varmus in 1998 was not initially well received by leaders of the ENT community,[142] since Battey is not an otorhinolaryngologist and his research in molecular biology had scant relationship to the research or clinical missions of the institute. He had started his NIH career as a senior staff fellow at the National Cancer Institute in 1983, served as a section chief in the Laboratory of Neurochemistry of the National Institute of Neurological Disorders and Stroke from 1988 to 1992, returned to the NCI until 1995, when he joined the NIDCD as director of intramural research.

Varmus arranged for Battey to be interviewed by a group of senior ENT surgeons:[153]

> As you'd expect, [Varmus wrote] when someone non-traditional is considered for a Directorship (this case was not unique, there is nearly always a worried faction), some of the ENT community was skeptical about Jim. On further exposure to him during the selection process, at least some of the leaders of that community came around. Since his appointment, Jim's been an excellent director and as far as I know there has been no further opposition or unhappiness. [Subjecting a potential institute director to such a review is] atypical but not unprecedented.[153]

The National Institute on Deafness and Other Communication Disorders conducts and funds research and training in a variety of areas consistent with its mission, such as: the normal and disordered processes of hearing, balance, smell, taste, voice, speech, and language, through a program of grants and contracts in basic, clinical, and translational research. The majority, about 55 percent, of NIDCD's budget is spent on hearing, about 6 percent on balance, 17 percent on smell, and 7 percent on taste.[151]

---

* M.D., Ph.D., resident in pediatrics, Stanford; postdoctoral fellow, Harvard.

Cochlear Implants

Among the advances Battey and the institute are proudest of is the cochlear implant for sensorineural hearing loss or "nerve deafness." NIDCD has supported research in this field since the early 1970s through its extramural and intramural programs.

Cochlear implants are used to improve the hearing of patients with malfunction of the "hair cells," tiny structures in the cochlea* of the inner ear that transmit sound signals to the auditory nerve, which conducts the information to the brain. The cochlear implant consists of an array of microelectrodes that is inserted surgically into the cochlea to stimulate the auditory nerve, which the damaged hair cells can no longer do normally. The patient wears an external computer-powered speech processor that detects sound and converts the sound into coded signals, which it then transmits through the skin to an implanted receiver that processes the signals and sends them to the microelectrodes in the malfunctioning cochlea.

"The cochlear implant is truly remarkable, a stunning achievement," said Battey. "A deaf child who receives a unit in the first year of life can be 'mainstreamed.'" The cost, however, is not trivial: $20,000 for the device and its insertion, and then about $35,000 over several months for otologists to teach the subject how to interpret the sound. "The socioeconomic status of the family is the leading criteria for success," Battey explains. "Being poor is now the biggest negative predictor."[151]

The institute also supports research to define how hearing loss is inherited. "With genetic counseling," said Battey, "we can tell a family the chances that their child will inherit deafness and relieve their guilt that something that happened during pregnancy caused it."[151]

National Human Genome Research Institute (NHGRI)

Established in 1989

2008 budget (in millions): $489

Intramural program: 20 tenured senior investigators, nine tenure-track investigators

---

* *Cochlea* is the Latin for "snail," which approximates the coiled appearance of the structure. The cochlea is about the size of a pea.

One of the NIH programs that has attracted the most notice in recent years is the Human Genome Project, whose aim was to sequence and understand the information preserved in the DNA of each human cell.

DNA is a chain of the chemical deoxyribonucleic acid in a specific sequence that defines its particular function. DNA consists of nucleotides, which are chemical combinations of sugars, phosphates, and bases,* and it is the sequence of the four bases that encodes the genetic information. Most, but not all, of each person's DNA is stored in the genes that encompass much of the information that leads to the formation and function of living organisms.

Humans genes are located on the 23 pairs of chromosomes in the nucleus, a small spherical structure within a cell that contains much of the cell's DNA. Some genetic material is also located in the mitochondria, structures within cells that convert food into energy. There are about 20,500 genes in each human, and the collection of the genes is called the "genome."†

Most of what genes do is not yet understood. Producing a single protein, probably their best-known function, involves only about one and a half to two percent of the genome.[154] Proteins consist of amino acids—small molecules that the organism must ingest or synthesize—in specific three-dimensional forms. Proteins carry out the duties within the cells that have been specified by the information encoded in the genes.

The structure of DNA, reported by James Watson and Francis Crick in 1953, and learning how the information stored in the gene produces a protein are among the most important discoveries in biology. The initial goal of the Human Genome Project, which was completed in 2003, was to create a "map" or "sequence" of the human genome that could be used to locate the genes within their chromosomes.

Since then, the institute that led the Human Genome Project, the National Human Genome Research Institute, has concentrated on understanding the one percent difference in the DNA of humans— 99 percent of the genome is the same in everyone—that accounts for, among other properties, our susceptibility to disease. Understanding the genetics of common diseases should, in time, allow medicine to become more personalized.[154]

---

* The bases in humans are the small molecules adenine, guanine, thymine, and cytosine.
† The word "genome" is a combination of the words "gene" and "chromosome."

Directors

NIH's involvement in what would ultimately become the Human Genome Project began in 1988 when James Wyngaarden, the director of the NIH at the time, called a meeting of scientists, administrators, and science policy experts to plan the effort. In October, the Office of Human Genome Research was created under the NIH director, and the NIH and the Department of Energy agreed to coordinate the work between them. A year later the Office became the National Center for Human Genome Research (NCHGR) with a budget of $50 million. In 2008 it would be $489 million, still only 1.6 percent of the NIH budget.

James Watson, who was appointed to plan[155] and later direct the project, predicted the work would take 15 years. "Twenty years was too long for Congress," Francis Collins, the director from 1993 to 2008, remembers being told. "Ten years was too few to do the work. Jim invented the 15."[12]

In April, 1992, 18 months after work began, Watson resigned. Bernadine Healy, then the NIH director, had prompted a review of Watson's investments, which showed that he and members of his immediate family had financial interests in several pharmaceutical and biotechnology companies and that he hadn't kept proper records of some of these associations. This, plus Watson's disapproval of some of Healy's policies, led to his resignation in April of 1992, an action that did not displease Healy.*

"She decided he had to go," said Francis Collins, Watson's successor, "and within 24 hours, he was gone."[12] The Healy vs. Watson imbroglio became a widely reported event in newspapers[158] and the scientific press.[156,159] "The fight with Watson was stupid," said former *Science* magazine journalist Barbara Culliton. "It didn't do Bernadine any good. It reinforced the notion that she was arrogant, not that Watson isn't also."[160†]

Michael Gottesman, a distinguished NIH molecular biologist and later the director of its intramural research program, became the

---

* Soon after leaving the NIH, Watson said, "I don't know how to get anyone to succeed me. I don't know anyone who doesn't have stocks. And I don't know anyone who would want to live with my boss."[156] Healy retorted, "Watson—the rules are not for him."[157]
† Healy and Culliton were classmates at Vassar College.[160]

acting director of the "infant that was kicking the slats of its cradle" as Collins describes the nascent project at the time.[12] A search committee got to work. Collins,[*] then leading one of the six genome centers based at medical schools, in his case at the University of Michigan, was contacted by one of the institute directors, who told him, "Bernadine wants you to apply." He did—"rude not to do so," he explained—but when offered the job, said no. "Our genome center was growing, I was a Howard Hughes investigator[†] and couldn't imagine moving my team of 12 investigators. Besides, I had no desire to get involved in the bureaucracy of a federal appointment." Despite these reservations, Collins eventually succumbed to Healy's entreaties as he realized that as director of the project he would have "the opportunity to design the direction of all the research" and wouldn't get so bogged down in administration that he couldn't continue to do science.[161] "I could make it happen," he concluded.[12]

In April, 1993, Francis Collins became the second director of the National Center for Human Genome Research, which later became the National Human Genome Research Institute. Donna Shalala phoned Collins from the University of Wisconsin, whose chancellorship she was resigning to become Secretary of the Department of Health and Human Services in the Clinton administration, to say, "It's great that we're both taking these jobs."[12] Four years later, Shalala supported the center's becoming one of the 20 institutes of the NIH at the time, thereby, among other perks, allowing the institute director to award funds of up to $50,000 without receiving approval from the institute's council.[12]

In time, the institute developed a program, both intramurally and extramurally, that Nobel laureate David Baltimore described as "on a scale never thought possible."[162] The institute employs about 430 people, of whom 30 are independent investigators. "All of us feel that it's a magical moment in science. It's transforming medicine," said Eric Lander, a leading genome investigator at MIT and Harvard. As for the genome institute, "it's *sui generis*."[161]

---

[*] Ph.D., Yale; M.D., North Carolina; intern and resident, North Carolina Memorial Hospital, Chapel Hill; fellow in human genetics and pediatrics, Yale.
[†] The Howard Hughes Foundation funds people. Except for the NIH Pioneer Awards, (see Chapter 9), the NIH funds projects rather than people.[100]

One-Hundred-Hour Week

As director of the NHGRI, Francis Collins, a member of both the National Academy of Sciences and the Institute of Medicine, still worked the schedule of an intern. At 57 he managed to spend 80 to 100 hours per week on the job, supervising the research in his laboratory—"all post docs, champing to be independent"—as well as leading the institute. The director is responsible for projects in 20 centers outside the NIH in six countries, some of which fund their work. "The job is incredibly demanding. It's so hands-on," Collins said. Concerning his schedule, he added, "other directors are much more sensible about this than I am."[12]

Collins Leaves

Francis Collins resigned his position as director of NHGRI in August, 2008.[163] He explained:[164]

> I am leaving because the time seems right for me to explore a number of professional opportunities in public and private sectors that I could not really even consider—and certainly not discuss or pursue—if I continued in my role as a federal employee. The rules would simply not permit it.
> In addition, I have writing projects in mind that I would like to pursue, especially about the future of personalized medicine—again, activities that I could not pursue while serving as director of the National Human Genome Research Institute.
> So I am going to take a kind of sabbatical for a few months—to write, to reflect, to spend some time trying to identify the next opportunity for service. I am not being coy with you—I do not have a definite plan of what that next step will be.
> I am not leaving because of any problems or disagreements with NIH leadership. Certainly the budgetary constraints on NIH have been deeply troublesome for the last five years—and that's something that I hope can be redressed in the future—but that is not the reason I am leaving. Despite these difficult constraints, NHGRI continues to conduct incredible research—perhaps just not as much as the opportunities would allow.

Zerhouni appointed Alan Guttmacher, deputy director of NHGRI, as acting director of the institute.

> **National Institute of Biomedical Imaging and Bioengineering (NIBIB)**
>
> Established in 2000
>
> 2008 budget (in millions): $300
>
> Intramural program: one tenured senior investigator, no tenure-track investigators

Founding the most recently established institute at the NIH, like the beginnings of the National Institute on Deafness and Other Communication Disorders, provides a story typical of how many of the other institutes and centers came to be and emphasizes the close relationship of the NIH to Congress and the Administration. What is unusual in this case is that the principal advocates were doctors and scientists rather than patients, members of their families, or disease-oriented not-for-profit organizations.

Creating the Institute

If Stanley Baum[*] and his colleagues in the Academy of Radiology Research had had their wish, a radiology institute would have been launched at the NIH long before 2000. "We tried for 20 years, but NIH didn't want it," he said. "Every institute does imaging research, and the directors saw a new institute [as] taking imaging away from them."[165] Accordingly, radiologists doing research in imaging had to apply to the existing institutes for disease-specific grants—such as the institutes of cancer, mental health, neurological disorders and stroke, drug abuse, and aging—and there was no locus for grants supporting basic research in the field. "Most of the fundamental research in our field was coming from other countries, particularly Britain and Sweden," Baum said. "That's where CT and MRI were developed."[165]

In addition to the institute directors with research programs in imaging, Harold Varmus, while NIH director, also opposed the creation of an imaging institute.[17] "Harold believed that imaging research

---

[*] Baum was chairman of the department of radiology at the University of Pennsylvania School of Medicine from 1975 to 1996.

was engineering and belonged in the National Science Foundation," said Baum.*

Baum and his colleagues in radiology joined with leaders in the bioengineering community, hired a lobbyist, and took on Congress. In the Senate, Trent Lott, encouraged about the project by a Georgia neighbor and bridge partner who was a radiologist, agreed to sponsor a bill to create the institute. Lott had become impressed with the effect technology was having on medicine. His friend, who formerly was called away from their dinner or card game to read a film at the hospital, could now interpret the image on his home computer, which was linked electronically to the hospital's radiology department.

When Lott tried to attach the authorization creating the imaging institute to a budget bill, Baum recounts, the committee chairman excised it on December 25, 1999. Lott then introduced a separate bill, which passed the Senate, as had a version in the House, "in the final moments of the last session of the Senate," wrote Varmus. "It was introduced on a point of order that received unanimous consent."[20]

The NIH adhered to its long-standing policy of opposing any further increase in the number of institutes and tried to convince President Clinton, who was disposed not to sign it since it was Lott's bill, that he should pocket-veto it. Lott told the NIH that if they wanted a new reauthorization bill, they should stop opposing the radiology institute bill. The NIH withdrew its objections, and the president signed the bill, one of the last of his administration, on December 29, 1999.[165]

Harold Varmus, who as NIH director and afterwards has championed an NIH with fewer institutes, wrote in 2001:[20]

> At no point were congressional hearings or public debates held to consider the possible effects of the new institute: How will the creation of NIBIB affect the many bioengineering and imaging programs now conducted through the disease-specific divisions of existing institutes? How will support for NIBIB affect the budgets of other federal science agencies that fund work in these areas but have fared less well than the NIH in recent years? Will the founding of the NIBIB promote or reduce the fruitful interactions among bioengineers, clinical investigators, and laboratory scientists that have been growing stronger in recent years?

---

* Varmus was heard to say about having a radiology institute at the NIH, "over my dead body."

Projects

The National Institute of Biomedical Imaging and Bioengineering is a non-categorical institute, like the National Human Genome Research Institute and the National Institute of General Medical Sciences. It is not based on an organ system (National Heart, Lung, and Blood Institute), on a stage of life (National Institute on Aging), or on a disease (National Cancer Institute). The first and current director is Roderic Pettigrew, a radiologist with an M.D. from the University of Miami and a Ph.D. in applied radiation physics from the Massachusetts Institute of Technology.

The intramural program at NIBIB is just getting started. The institute has recruited investigators working in imaging from other institutes, but only three percent of the budget is spent intramurally. Consequently, a larger portion of the institute's budget is allocated to the extramural program than is typical in the other institutes.*

Among the projects the institute currently supports are:

- Contact lenses that, in addition to their optical purposes, will measure the level of glucose in tears.
- An implantable device that will produce regeneration of brain tissue damaged by stroke.
- A computer-driven device to assist paralyzed patients with preserved cognitive function to convert their thoughts into physical actions.
- Methods to reduce the amount of time now needed to treat cardiac arrhythmias with catheter ablation.
- Techniques to more rapidly diagnose clinical problems, such as determining the DNA of bacteria causing urinary tract infections from a single drop of urine.
- An imaging system to analyze diseased tissues in the body that might eventually replace the traditional technique of studying specimens removed by surgical biopsy.
- Improved methods for conducting hemodialysis for patients with kidney disease.

NIBIB also provides support for innovative training using interdisciplinary techniques in topics related to the institute's missions.

---

*The research that Elias Zerhouni conducted at Hopkins before he came to the NIH might now be funded by NIBIB.

CHAPTER FOUR

The Centers

IN ADDITION TO the 20 institutes, there are seven centers, five of which serve administrative and clinical functions of the National Institutes of Health.

Center for Scientific Review (CSR)[1]

Established in 1946

2008 budget (in millions, taxed from institutes): $100

Intramural program: None

The Center for Scientific Review[*] receives and processes each of 75,000 grant applications sent to the NIH annually at a cost of about 0.2 percent of the amount requested.[2,3,5] CSR reviews and evaluates about three-fourths of them. The others are evaluated in the institutes and other centers.

---

[*] Formerly the "Division of Research Grants" (an institutional promotion, since a "center" is higher in the NIH pecking order than a "division"). Outside the NIH, CSR is often called "Submission and Assignment."[2]

The annual CSR budget is about $100 million: $60 million for operating the CSR and $40 million for travel, hotel, food, and a small honorarium for the 18,000 reviewers who participate in the peer-review process.[6]

Study Sections

Referral officers assign each application to an Integrated Review Group (IRG) within CSR or to an institute or center for potential funding. The more than 20 IRGs are grouped into four divisions:

- Biologic basis of disease
- Molecular and cellular mechanisms
- Physiology and pathology
- Clinical and population-based studies.

An example of an IRG is "Cardiovascular Sciences (CVS)" in the division of physiology and pathology. Each application that CSR receives that is best reviewed in this IRG is assigned to one of its 11 study sections, which have names (and acronyms) like: "atherosclerosis and inflammation of the cardiovascular system (AICS)," "hypertension and microcirculation (HM)," or "myocardial ischemia and metabolism (MIM)." In the summer of 2008, the NIH operated about 200 study sections in CSR and additional review groups in the institutes and other centers.[2]

The NIH scientific review officer (SRO)* who is assigned to this application decides which study section can best evaluate it. About six weeks before the study section meets, the SRO sends all the applications to be reviewed at this session to the members of the study section to study. Three members of the study section, at least two of whom will provide written reviews, examine the application assigned to them in particular detail. Before the study section meets, reviewers submit confidential preliminary critiques and scores to CSR. Applications scored in the lower half are not discussed at the meeting.

Study sections—most have about 20 members—meet three times per year, usually for two very busy days. Although each member has received copies of each application to study in advance of the meeting, the three members assigned to the particular application being

---

*Formerly known as "scientific review administrator" and "executive secretary."[2]

considered will lead the review. After general discussion, the members of the study section register their scores privately. The CSR staff tabulate an average, and this becomes the "priority score" for the application.

Within a few days after the meeting, applicants receive their priority scores and percentile ranking electronically. In about a month, summary statements of the study section's deliberations are also transmitted and include:

- Critiques written by the assigned reviewers
- Summary of the discussion at the study section, prepared by the scientific review administrator*
- Recommendations of the study section
- Administrative notes for special consideration.

When is the best time during the year to submit a grant application many an applicant has wondered. "It should be in principle," answers CSR Director Antonio ("Toni") Scarpa, "but it is so unpredictable that one would need a very efficient crystal ball to figure out when."[6]

The topic of the research, however, can help determine whether a high-quality application gets funded. If Nobel Prize winners, for example, are working on it, the chances of funding may be less than when less-renowned investigators are competing. Applicants who work in less fashionable projects can win funding while those studying more popular subjects may have less success.

Study Section Members

Membership on a study section means that the member is an authority in one or more areas of biomedical research, and this membership is a task requiring intense study by investigators whose days are already full.[7] Many permanent members, who serve for terms of four years, are chosen based on their performance as *ad hoc* reviewers on study sections requiring their particular expertise.† As the number of applications in particular subjects grows, CSR adds study sections and members to keep the work of each section within workable limits.

---

* These are some of the people at study sections who sit in chairs against the walls.[5]
† The NIH publishes specific criteria for the selection of members of study sections[7] and the responsibilities of study section chairs.[8]

The term limits of study section members mean that when an application that was returned to the author for revision reappears at the NIH, the original reviewer may have left. Although a detailed appraisal of the original application is available, the new reviewer will not be as informed about the application as the original reviewer would have been. An extramural and former intramural scientist regrets that the staff in some cases may not give adequate help to the reviewers in understanding what the previous study section thought about the application.[9] "There can be disadvantages," said Toni Scarpa. "When reviewers assess an application a second time, there is a tendency to give better scores for minor improvements that a new reviewer would not find as significant."[10]

Convincing leading medical scientists that they should serve on study sections can challenge the persuasive skill of administrators in CSR and the institutes. "Sometimes, I have to call 20 people before I get someone," says Suzanne Fisher, director of the Division of Receipt and Referral since 1996.[2]

The increasing complexity of biomedical research complicates the work of the NIH administrators, who are responsible for staffing the study sections. "It's difficult to find experienced basic scientists who are also familiar with research in clinical subjects, or clinical investigators who understand the intricacies of basic issues like genetics, for example," said Stephen James from NIDDK. "Many basic people see disease-oriented problems as peripheral to their interests and may even discount the value of clinical research."[11]

And there are the practical limitations. "Busy people just don't want to travel, with all the plane cancellations and other harassments these days, and the trip to Washington and then out to Bethesda can be quite a trial," said James. "I've heard that the NIH is a major user of hotel space in D.C."[11] Although the NIH conducts some reviews electronically, this method precludes the face-to-face contact of the peer reviewers whose verbal deliberations refine the relative merits of the applications being evaluated.

Doctors and scientists are not known for writing clearly, so miscommunication can interfere with an application's receiving the most advantageous review. In the past, larger programs were reviewed by "site visits" in which a peer review group would visit the institution that submitted the grant, tour the facilities, and listen to presentations by the investigators, who could try to clarify orally the inadequately explained parts of their applications. Sometimes the investigators, as a group, would come to Bethesda for a similar review, called a "reverse site visit."[2] Both of these techniques to refine the merits of applications have, for the most part, been abandoned because of the time, labor, and expense required for the many people involved and the limitations that keep many potential peer reviewers from accepting such assignments.[11]

Trying to make service as a reviewer easier is an important goal of Toni Scarpa, who has directed CSR since 2005.[*] In particular, he wants to find better ways to recruit more senior, established investigators for the study sections.[12] They could serve only one or two rather than three sessions per year and convey their opinions online or by telephone, thereby avoiding the anguish inherent in traveling.[13]

Peer-Review System

"Our peer-review system is the glory of the NIH when compared with the way other countries do it where there's much more decision-making by senior people,"[14] said Steven Hyman, former director of the National Institute of Mental Health. Antonio Scarpa agrees. In Europe, said Scarpa, who was born and educated in Italy, "only five to ten percent of all the money spent on biomedical research is peer-reviewed. The government pays faculty salaries and assigns research support, and the university has extra money to support research." Except for the F. Edward Hébert School of Medicine of the Uniformed Services University of the Health Sciences, located in Bethesda across Rockville Pike from the NIH, "the United States government has no commitment to any medical school or faculty," said Scarpa.[5]

The peer-review system of the NIH is thought to constitute "by far"[15] the best system for allocating money for research and to have contributed significantly to the preeminence of American biomedical research.[15] "It's largely responsible for the growth in medical science," said Nobel Prize winner Michael Brown, "and supports science for its own sake avoiding things that are politically correct. Peer review works fabulously when there's lots of money, but less successfully when dollars are scarce. Since just one reviewer saying something negative can kill an application, only the bland ones get through."[16]

---

[*] Scarpa was chairman of the department of physiology at the Case Western Reserve University School of Medicine when appointed to the NIH.

The system has its critics, particularly when funding is limited, as in recent years.[17-20]* Brown is not alone in complaining that peer review at the NIH is too conservative,[9,19,22-26] that it "stifles innovation,"[23] "doesn't award high-risk research,"[9] and prevents people from "thinking out[side] of the box."[22] Submissions without much data, even those with challenging ideas, are infrequently funded.[26]

Richard Klausner director of the National Cancer Institute from 1995 to 2001, believes that peer review, though "absolutely central to a healthy approach to science," overemphasizes the technical aspects of the proposals it reviews. "It's overwhelmingly concerned with feasibility and misses the forest for the trees."[27] Klausner suggests that the intense competition for support has tended to produce "small projects that are more and more reductionist. You see a lot of 'me-too' science, work that is very safe and repetitive."[27] Consequently, he likes many aspects of Elias Zerhouni's Roadmap, particularly the Pioneer awards that support research that is creative in nontraditional subjects.

Klausner has found that reviewing for foundations can be more satisfying than serving on NIH study sections because the criteria for awarding may be more innovative.[27] In talking about the NIH this way, Klausner knows that he is criticizing "a sacred cow. People say these things in private, not in public, because they're concerned that such comments may discourage the public and Congress from supporting NIH."[27]

Vincent DeVita, like Klausner a former NCI director (1980–1988),† suggests that "the peer-review system is a religion. It is troubled, and anyone who criticizes it is Public Enemy Number One."[28] DeVita disapproves of the fact that the investigator-initiated R01 grants dominate the NIH's extramural awards. "The NIH has become a slave to them. We found [in the NCI] that the R01 is not the type of grant that supports most important discoveries. Goal-directed grants and contracts are often as valuable and more flexible. The leading investigator can reassign parts of them as the progress of the research dictates." In DeVita's experience, "R01s don't work for clinical research."[29] He adds that "study sections can be influenced by non-scientific matters."[28]

Nobel laureate David Baltimore says, "The peers are not peers. There's too much attention paid to getting a variety of people on the

---

* As Michael Lauer of the NHLBI, who was previously supported in the extramural program, said, "I love peer review, when they fund me."[21]
† DeVita directed the Yale Cancer Center from 1993 to 2003 and is now chairman of the Yale Cancer Center Advisory Board.

study sections rather than real scientists."³⁰ Harvard investigator Marc Kirschner puts it more specifically: "Diversity may not help."¹³

Baltimore believes that, since the volume of applications is so large, "efficiency and effectiveness are low. The reviewers are called on to make very fine judgments with too little money."³⁰ Senior members with strong reputations can "take control of the room" and dominate the result, despite different opinions from younger and less distinguished members.²¹

"Research moves faster than the study sections," says Kirschner. The subject covered by a particular study section may become less fashionable and less challenging. When this happens, "mediocre grants get funded," while applications assigned to a study section covering the most contemporary and competitive subjects may miss being funded even though the work is of higher quality and more creative.¹³ "Many people are uncomfortable with research that takes risks or is novel," says Kirschner.¹³ "Its [the system's] biggest problem is that we pay for work that's already done," is another complaint.³¹

Others criticize the peer-review process for letting ethical considerations cloud its impartiality. Adil Shamoo, a basic science professor at the University of Maryland and an authority on conflicts of interest in research, believes, "the peer review system is fundamentally corrupt."³² The grants are reviewed, Shamoo explains, by the applicant's competitors. "If they fund you, there's less money for me," he imagines many of the reviewers thinking.³³ In the process's current form, the advice of the primary reviewer and often, to a lesser extent, the secondary reviewer "carries the day. Consequently only one or two people decide the funding likelihood of a grant." The other members of the study section are expected to study each of the grants being reviewed so that they can offer an independent judgment. All too often, Shamoo says, "reviewers do this superficially."³³

Shamoo has advised the NIH to engage scientists working in fields other than those of the applicants and then retain advisors to explain and judge the accuracy of the technicalities of the research. "No study section can pick up innovative ideas," he believes. "They only know what's good now and give block grants based on the career of the applicant. Scientific review officers have tremendous power. They're the ones who choose the reviewers."³³ Scarpa replies, "This is not the case at CSR, where scientific review officers in the institutes and centers have the primary responsibility for recruiting their reviewers."¹⁰

Norka Ruiz Bravo, deputy director for extramural research, recognizes that study section members may have conflicts, "but NIH," she says, "has a robust system for ensuring that those with conflicts do not participate in the reviews of applications for which they have conflicts."[34] Others agree. "I think it works better than many people outside the NIH believe," says Anthony Demsey, a senior administrator in the National Institute of Biomedical Imaging and Bioengineering (NIBIB).[12] Reviewers must recuse themselves from reviewing any application with which they may have a conflict of interest. "Conflicts are understandable," Tony Fauci admits, "but fixable."[35]

Reviews by the Institutes and Centers; Division and Program Directors

The institutes and centers that award grants review about one-fourth of the applications that come to the NIH. These consist of applications specific to particular institutes, such as training grants, program project grants, large clinical trials, consortia, networks, and research centers.[2,36] Most requests for applications are also reviewed at the institutes unless CSR can do them more conveniently.[36] Although the applications reviewed in the institutes constitute a minority of those sent to the NIH, they receive a larger proportion of the extramural budget since the programs they review tend to be larger than those reviewed in the Center for Scientific Review.[2]

Among the senior institute administrators who supervise this process is gastroenterologist Stephen James, director of the division of digestive diseases and nutrition, which includes obesity, in the National Institute of Diabetes and Digestive and Kidney Diseases (NIDDK).* Two other division directors handle submissions to study the other diseases assigned to the institute.

About 20 program directors work in Steve James's division, each responsible for one or more areas of research.[37] Since James and the program directors in his division "want the best science, we try not to direct money to any particular area."[36] Nevertheless, program directors are occasionally accused of bias about which grants reviewed in the

---

* Disclosure: Stephen James was formerly a colleague of mine at the University of Maryland School of Medicine, where he was head of our gastroenterology division from 1991 to 2001.

institutes get funded. "It's true that we can make or break a review by picking the 'right' people for our study sections," Steve James says. "CSR may do a fairer job since it is designed to keep reviews autonomous and away from the prejudices of the institutes."[36]

The program directors and their staffs review every application that the study sections decide warrant support, plus some receiving scores just below the "pay-lines." The institutes can fund such grants if their staff thinks that an application has particular merit or promise that the members of the study section may not have appreciated.* In the past, administrators in NIDDK have kept about 10 percent of their budget to support initiatives of particular interest. "Flat budgets have, in effect, killed these programs," says James.[36]

Ironically, support for research does not necessarily correlate with the magnitude of the disease in the population. "In gastroenterology," explains Steve James, "gall bladder disease is much more common than many of the GI diseases that are now research-fashionable, and the surgical treatment is excellent, but there's little research into its cause these days."[36]

Program directors and their staffs in the institutes and centers are responsible for assuring that each application fulfills the criteria for the care given to each human and animal subject as determined by each institution's institutional review boards, and determining whether there are any potential legal problems.

The program directors and the scientific review officers, whom extramurally funded investigators contact at the NIH, can become the victims of the disappointment applicants feel when their grants are not funded. The program directors are criticized for being administrators, which they are, and not working scientists,[31] which, despite the opinion of applicants, many are or have been.[39,40] "Are you telling me that a failed scientist is deciding about my grant?" many a disappointed applicant has complained.[41]

To maintain impartiality between how the program directors in the institutes and centers think about a grant versus the opinions of the officers in CSR, members of one group do not discuss the merits of particular grants or programs with the other. This prevents the particular enthusiasms of program directors in the institutes and centers from influencing the scores assembled by the CSR administrators and

---

* Funding grants that are out of order—"below the payline"—aroused the attention of Senator Grassley (R-Iowa) who charged that one of the institutes had inadequately documented the reasons for supporting such grants.[38]

vice versa.⁴² The NIH emphasizes that a wall must exist between the program directors who develop research agendas in the institutes and centers and the scientific review officers who administer the reviews, so that the decisions of one do not affect the decisions of the other.

Early in his tenure as NIH director, Harold Varmus was heard to say about the program directors, "They're bureaucrats, who don't know [much] about science," although he used a somewhat more pungent word to describe their knowledge. "We changed his mind," says one of the institute directors. "He left with a better understanding of what these folks do."

Stephen Katz, director of the National Institute of Arthritis, Musculoskeletal, and Skin Diseases, praises these essential members of the NIH administration. "We say that our program directors are not just deliverers of pink sheets [the reviews of study sections sent to applicants before electronic reporting replaced them], but real scientist-administrators and strong advocates for science." In his institute, Katz expects these administrators to attend scientific meetings as well as study sections and talk to applicants "with truth and knowledge, which includes telling them whether their applications are fixable or not."⁴³

The NIH leadership is aware of the plight of the 3,000 program directors who have been stigmatized as "paper-pushers." There used to be a consensus that many had failed as intramural scientists and were transferred to the extramural program. Contrary to what many think, some program directors are respected scientists who have chosen a different career path.

Review of the Peer-Review System

Although one former institute director believes that the NIH director would be "dumped on if he tried to change the peer-review system,"²⁸ Elias Zerhouni decided to respond to criticisms of the system. In June, 2007, he announced the formation of internal and external working groups* "to examine the NIH peer-review process, with the goal of

---

*The external committee, made up of leading investigators working outside the NIH, was co-chaired by Keith Yamamoto, executive vice dean and former chair of the department of cellular and molecular pharmacology at the University of California–San Francisco, and Lawrence Tabak, director of the National Institute of Dental and Craniofacial Research, who acted as a liaison to the internal committee of leading NIH scientists and directors, which he co-chaired with Jeremy Berg, director of the National Institute of General Medical Science.¹⁵

maximizing its effectiveness,"[44] and asked for the advice of all interested people.[45] Although many at the NIH felt that changes were necessary, there was no consensus about what changes should be made, other than that the process was too conservative.[22] "It's not broken but it needs repair," in the opinion of Keith Yamamoto, who chaired one of the review committees.[46]

The "diagnostic" phase of the study, which was completed by the end of 2007, was presented to the full Advisory Committee to the Director in December and posted on the NIH website on February 29, 2008.[47] The "implementation" phase began in March 2008.[15] Some of the recommendations were:[46,48–50]

1. Applications and reviews
   a. Shorten length of R01 applications from 25 pages to 12 pages.
   b. Reviewers should focus more on the anticipated impact of the research and less on methods and other details. This should help "recapture the spirit of study sections in the old days," co-chairman Lawrence Tabak said.[48]
   c. Reviewers will score on five criteria, in addition to an overall score to provide clearer advice to applicants.
   d. All applicants will receive scores.
   e. At the end of a study section meeting, reviewers will rank applications.
2. Reviewers
   a. Reviewers can participate in 12 sessions over six years instead of four years and possibly share the duty with a colleague.
   b. NIH will offer reviewers who have participated in at least 18 study section meetings the opportunity to apply for administrative supplements of as much as $250,000.
   c. Those receiving high-prestige awards from NIH or holding at least three basic research grants will be obliged to serve if asked.

Responding to criticisms, mostly from members of the extramural community, NIH officials rejected some of the recommendations; for example:

- Consider all proposals "new" so that resubmitted ones get no particular advantage. Instead, NIH plans to investigate ways to fund a larger portion on the first round.
- Fatally flawed applications should be labeled "not recommended for resubmission."

- To reduce the size of the overhead payments, institutions should assume more financial responsibility for building and operating laboratories and for investigators' salaries.

Computerized Applications[51]

In December, 2005, the NIH began replacing paper applications with electronic submission of applications, which by the fall of 2008 was 80 percent complete.[52] When fully instituted, the electronic system should save about $10 million per year.[53]

Applicants send their completed forms to the website "Grants.gov," which is based in the Department of Health and Human Services and will eventually receive grant applications for all agencies of the government.[5,54] All computerized grant applications must be written on PCs; the system does not accept material from Apple computers.[5]

About 1,000 federal programs, which disburse more than $400 billion annually, are coordinated though the Grants.gov website.[55] The Electronic Research Administration (eRA) at the NIH receives its applications as Adobe files from the DHHS office of research. After checking them for compliance with human and animal welfare, eRA forwards the applications to the Center for Scientific Review for further processing.[5,54]

When the system is fully functioning, CSR should be able to operate with 20 to 30 fewer employees—the center employed about 300 in the summer of 2008.[10] Processing a computerized application, however, takes about the same amount of time as an application on paper does.[5] What is avoided, of course, is the physical handling and storage of the 75,000 paper applications per year and the cost to both NIH and applicants of mailing them.[2]

Each applicant registers with "eRA Commons," which becomes the electronic communication link between the applicant and the NIH for all news about the application. Through the "Commons account," the applicant learns how the study section evaluated and graded the grant more quickly than when this information was transmitted by mail in the past.[5] The Electronic Research Administration is the office that authorizes the government disbursing agency to send the money to successful applicants and their institutions.[54]

In 2005, the Office of Management and Budget announced plans to consolidate the 26 federal agencies that offer grants to about five "service providers." The NIH was one of those selected, and it now provides electronic services to, in addition to itself, the Centers for Disease Control and Prevention, Substance Abuse and Mental Health Services Administration, Food and Drug Administration, and Agency for Healthcare Research and Quality. Not all of these agencies were particularly enthusiastic about being grouped together for this service under the NIH.[54]

NIH Clinical Center (CC)

Established in 1953

2008 budget (in millions, taxed from institutes): $352

Intramural program: 20 tenured senior investigators, two tenure-track investigators

---

The Clinical Center is the NIH's hospital on the Bethesda campus.[56] Like all patents in the center, its first patient, who was admitted in 1953, came there to participate in clinical research being conducted by scientists in the institutes. The original building, later named for Senator Warren Magnuson, a particularly enthusiastic supporter of the NIH, had 500 beds, but in recent years, many fewer beds were filled on most days.[57] Since then, the Clinical Center has added more buildings, the latest being the Mark O. Hatfield[*] Clinical Research Center, a 242-bed replacement hospital opened in 2005 and attached to the older Magnuson building. Congress initially appropriated $320 million for the new hospital. The final cost was $500 million. To pay for the overrun, construction of half of a new neurosciences research building was postponed.[57] Because there continue to be unfilled beds, John Gallin, the director, wants "to open the doors to the extramural community."[57]

All inpatients are now treated in the Hatfield Center. Ambulatory (outpatient) care, clinical laboratories, radiology, and the surgical operating room remain in the older buildings.[†] On the NIH campus near the

---

[*] Many of the NIH buildings are named for senators and representatives, often chairs of relevant committees and subcommittees, whose influence has brought the NIH the funds needed for their construction.[58]

[†] There are 8,000 inpatient admissions per year; average hospital stay, 8.8 days; in-patient census, 60%–70% of 234 open beds; 85 one-day accommodations; 100,000 outpatient visits per year; staff follows about 85,000 patients. In the past decade, inpatient volume has increased by about 9%; outpatient, 20%. Staff: 1,200 physicians, dentists, and Ph.D. investigators; 660 nurses; 570 allied health-care professionals. Research: 1,600 laboratories conducting basic and clinical research; 1,375 active research protocols, about 25% of which change each year.[57,59,60]

Clinical Center are two temporary residences for families of patients receiving care there: the Edmond J. Safra Family Lodge for adults and the Children's Inn at NIH for children.

Few acutely ill patients come to the Clinical Center, and those who do are transferred to Suburban Hospital across Old Georgetown Road behind the NIH or to Bethesda Naval Hospital across Rockville Pike.[61]*

The new Clinical Center has brought significant changes in its operations, predominantly due to the decreased number of inpatients in the new building compared with occupancy in the older building decades ago, and because of more outpatient visits. Today there are fewer patient care units and, whereas the institutes used to have their own wards with their own nursing staff regardless of how many beds were filled, today patient-care units are shared. Now patients from more than one institute may be admitted to the same nursing unit, in which the nurses may care for patients on different research protocols. In the past their work could be more specialized. NIH administrators talk about the decreasing number of "silos" in the Clinical Center as more governance has had to be shared.[62,63]

Patients

Each patient at the Clinical Center is a volunteer: "our partners or co-producers of our product," according to David Henderson, who is responsible for the physicians as deputy director of clinical care in the Clinical Center. "We're a high-risk, high-reward place," he said.[64]

Many patients, though fewer than in the past, are referred to the NIH by their physicians. About 40 percent refer themselves, often from learning through the internet that the NIH is studying their disease.[65]† The office of clinical recruitment in the Clinical Center or similar groups in the institutes review the suitability of every applicant to assure that they have "orphan diseases"—rare ailments that are not treated at most centers but are being studied at the NIH—or more

---

* One severely ill patient on the psychiatric service committed suicide six months after the new hospital opened. He managed to climb over the eight-foot glass barrier that surrounds the walkway on the eighth floor and fall to the central atrium below. A permanent barrier has been installed between the walkway and the atrium that extends to the ceiling on all floors.[62]

† Clinicaltrials.gov. "Many patients, particularly those with managed care contracts who seek admission to the Clinical Center," John Gallin has observed, "do not know who their primary care doctor is."[57]

common illnesses being investigated by scientists at the Clinical Center. About half participate in special studies related to their illnesses, and the others in trials of new drugs.[64] Healthy volunteers, fewer now than in the past, come to the CC to participate in Phase I and Phase II clinical drug trials.*

"We don't mimic what's being done on the outside," said one of the leaders in the Clinical Center.[66] Pharmaceutical houses have little interest in developing treatments for the rare diseases that afflict most patients in the Clinical Center. Because of their novelty, obtaining extramural support for the work that the Clinical Center investigators perform would be very difficult.[11] "It's something the government should support," said one of the NIH scientist-administrators.[11]

Although they come to the Clinical Center from most states, about 70 percent of the patients live in Washington, Virginia, Maryland, or other nearby states. Many will need care as outpatients, precluding commuting from many parts of the country. International patients infrequently come to the Clinical Center.[57]

The patients pay nothing for their care. The federal government pays all costs of their hospitalizations and ambulatory visits from the NIH budget. The government even covers the expenses for travel and lodging for them and the families during the studies.[67] The NIH receives no reimbursement from Medicare, Medicaid, or private insurance.[42]

"We don't have to worry about market share or a practice plan" said David Henderson in comparing the Clinical Center with other teaching hospitals and medical schools.[64]

Each room in the new hospital accommodates one patient and has a combination TV/computer through which the center transmits educational programs about medical diseases. Rather than clean them, the center uses disposable keyboards and mice. "We have a different mindset in this building from the rest of government," said Gallin. "We're doing something special. The patients love us better than anywhere else."[57]

---

* Phase I studies establish the chemical and metabolic effects of a new drug in a small group of human volunteers. Phase II studies test the drug for safety and efficacy in patients with the disease for which the drug was developed, or in volunteers. Phase III studies are larger trials that compare the drug with standard therapy currently being used for the disease in question. Phase IV studies include many subjects and are conducted after the drug has been approved by the Food and Drug Administration (FDA). These trials compare the drug with a competitor, explore additional patient populations, or further study adverse events.

Most of the day-to-day care is given by nurses, many with graduate degrees, who become knowledgeable in the research being conducted. With 96 percent of the 700 nurses being R.N.s., the Clinical Center employs many fewer practical nurses and nurses' aides as a fraction of the nursing personnel than do most hospitals. Some of the nurses are members of the Commissioned Corps of the Public Health Service, and consequently, may be assigned temporarily outside the Clinical Center when the PHS has special need for them elsewhere.[68]

"They come to work here to do something beyond themselves," said Clare Hastings,* chief of nursing and patient care services in the Clinical Center, when describing her staff. "Every patient has either a bizarre disease or needs specialized care, sometimes as a last resort. Some choose us for altruistic reasons. They want to provide information that could help other sick people."[64,67]

Research is facilitated by the proximity of the beds to many of the investigators' research laboratories, which are often across the floor from the patients,[69] "making 'bench to bedside' real," said Harvey Klein, a leading clinical investigator. "It's no problem if the patients stay weeks or months—though such prolonged admissions were more common in the past than now. We clinical investigators do not see money as a driving force inhibiting our work."[70] Although many clinical research projects at the NIH occur in the Clinical Center, experiments involving animals are conducted in the institutes and not in the Clinical Center.[66]

Since much of the research conducted in the Clinical Center originates in the institutes, Gallin and his associates have to negotiate with the institute leaders to achieve consensus on some matters.[71] Accordingly, the Clinical Center has been called "a hospital with 20 medical schools, each with its own dean and agendas, some competing with each other."[64†] Nevertheless, surgeon Steven Rosenberg said,

---

* B.A., Reed; B.S.N. and Ph.D., Maryland; M.S., Georgetown.
† Institutional review boards (IRBs) must review and approve the clinical research that all institutes perform. Each institute either has one IRB or, in the case of institutes with smaller programs, collaborates with IRBs in larger institutes. Some IRBs have recently been consolidated with other IRBs, a process that Director Zerhouni encourages.[72] "The process of consolidating the IRBs helps to reduce the possibility that institute leadership might exercise inappropriate influence over IRB decisions," said Jerry Menikoff, director of the Office of Human Studies Research.[73] At the end of 2008, there were 12 IRBs for the intramural program.[73]

The NIH's IRBs operate under a "common rule" for most governmental agencies that conduct research on human subjects.[72]

"Without the Clinical Center, the intramural program would not be unique. It's where we take the science we produce to the bedside."[74]

Management

In the 1980s and early 1990s, many considered the Clinical Center to be poorly managed with no financial acountability.[64] "One of the complainers was John Gallin," then scientific director for intramural research at the National Institute of Allergy and Infectious Diseases, remembers Stephen Schimpff, a senior executive with the University of Maryland Medical System at the time and soon to be a member of the NIH committee charged with improving the administration of the Clinical Center. "Accordingly, John was appointed director."[71]

Described as a "highly respected investigator," "very open and reactive," and "with a strong, clear vision that makes things happen,"[61,71] Gallin had started at the NIH as a trainee in 1971.* As director of the Clinical Center, he continues to lead a research laboratory studying the hereditary disorders of the cells that combat infection† and regularly makes rounds on patients with infectious diseases. "I stay here because I have a passion for the place," Gallin explains.[57]

Partly because of its history of unsatisfactory management, the NIH almost lost the right to directly run the Clinical Center. Soon after the Clinton administration took office, the Office of Management and Budget (OMB) pursued contracting the operation of the hospital to a teaching hospital or another company as part of Vice-President Al Gore's "Reinventing Government" initiative.[57,71,75] Gallin and Harold Varmus, the NIH director at the time, who thought outsourcing the management of the Clinical Center "a terrible idea,"[75] met with Donna Shalala, Secretary of the Department of Health and Human Services, and

---

* B.A., Amherst (like Harold Varmus who appointed him director of the Clinical Center); M.D., Cornell; medical training at Bellevue Hospital; research training at the NIH. When Varmus appointed Gallin, he charged him with focusing on building a new hospital and on training.[57]

† His research has long stressed the unusual illness, "chronic granulomatous disease," which, in view of its rarity, is a particularly suitable and representative subject for study at the NIH. "John has more than 100 patients with this disease," said Harvey Klein, chief of the department of transfusion medicine in the Clinical Center. "During eight years at Hopkins, I saw only one case."[66]

extracted a pledge from her to conduct a review of the Clinical Center before she decided whether to do as OMB wanted.

Shalala then created a "Clinical Center Options Team," led by Helen Smits,[75] an internist, who was then deputy administrator of the federal government's Health Care Financing Administration (HCFA).* Members of the group visited 30 health facilities and other relevant organizations to understand how to best operate the Clinical Center using the most contemporary methods of hospital management.[76] People at the NIH told Smits that "there is nothing we can learn from elsewhere because we're so different."[75]

What the committee found was that many of the Clinical Center's operational problems were the same that the leaders of other teaching and research hospitals were experiencing. Some issues were unique, however, such as the responsibility of the nursing staff at the Clinical Center to participate closely in the research. The group had heard that government regulations were so onerous that progress would be difficult. "Not so," said Helen Smits. "They were not nearly as limiting as we expected." Under the stimulus of the committee's advice and the threat of outsourcing its management, the NIH leaders found that they could surprisingly easily improve many of its customs and rules.[75]

The committee's report was issued in 1996, and among its recommendations were:[57,60,71,75-77]

- Build a new hospital.
- Institute specific measures to operate the hospital more efficiently. The hospital hired an expert in hospital management to "tutor us in holding people accountable," as Gallin describes his responsibility.[57] This led to the discharge of 78 underperforming employers in the course of three years. "Everybody said we couldn't do it, but we did."[57]
- Improve relationships with patients by surveying and then acting upon their advice about being treated in the hospital and the ambulatory clinics.
- Create a board of governors for the Clinical Center—now called the Advisory Board for Clinical Research—which would meet for the first time in 1996.

---

*Smits saw Shalala's plan for reforming the Clinical Center as a function of the Democrats' wanting to improve how government functioned, whereas the Republicans "wanted to get rid of it."[75]

- Create a budget specifically for the Clinical Center.
- Establish more educational and training programs.[78]
- Give the director of the Clinical Center more authority to make the hospital operate as a single organization rather than a collection of silos. This required reducing the independence of the institutes to manage their own floors, where many of the beds were empty.[*] The institutes, which were operating almost autonomously, should no longer make decisions affecting the Clinical Center on their own.[75]
- Collaborate more effectively with the extramural research community.
- Develop a strategic plan.

Gallin, Smits, and Varmus presented the study's advice to Shalala, who responded "We're going to do it all," and that is what has happened. Contracting the operation of the hospital to an entity outside the NIH was rejected.[57,75]

The budget

Formerly, the institute directors decided how much of their budgets to assign to clinical research in the Center. This practice produced inconsistent use of the Clinical Center and reduced the enthusiasm of investigators in some institutes for working there. Now, each institute with an intramural program supports the Clinical Center with funds equivalent to a specified fraction of its intramural budget—"like a school or library tax," Gallin explains.[57] One-third of these institute budgets is assigned to clinical research, and the Clinical Center receives one-third of that. As Michael Klag, dean of the Bloomberg School of Public Health at Johns Hopkins University and a member of the Clinical Center's advisory board for clinical research, puts it, "They pay up front whether they use it or not."[79]

This change has led the institutes to use the Clinical Center more often—"now they use it because they have to pay for it," said Gallin[57]—although the flat NIH budgets in recent years have reduced the money that the institutes can assign to clinical research. As recommended in the report, the Clinical Center may now carry forward reserves from year to year.[76]

---

[*] This will sound very familiar to all leaders at academic medical centers.

"John deals very creatively with the flat budgets and solves our financial issues better than do any of the institutes," said David Henderson. "Before the flat budgets, we paid little attention to costs. In the past, when we ran a deficit, we had to turn to the institutes at the end of fiscal year—very unpopular. Under John, this hasn't had to happen."[64]

Henderson[64] and his colleagues[11] acknowledge that the flat budgets have especially hurt the Clinical Center. "If you lose eight percent in a research lab, you close the lab for a month and write papers. Here, the inflationary pressures are tremendous."[64] The ten percent that is spent on drugs grows from six to fourteen percent per year. "Drug creep," Henderson calls it.[64] The government annually orders an inflationary increase in employees' salaries, which constitute 55% to 60% of the costs in the Clinical Center. With the budget not growing proportionaly, these costs have to be found in other parts of the budget, like firing under-performing employees over a three-year period.[57]

Despite the improvements in financial management, the Clinical Center remains an expensive enterprise. The fixed costs spread among relatively few patients remain high. High-tech equipment, such as the superb instruments the NIH uses in imaging, are very expensive and employed much less frequently than at most hospitals.[80] The PET (positron emission tomography) scanner, for example, is used, on average, only twice per day.[71]

Advisory Board for Clinical Research (ABCR).[57,77,81–83]

ABCR counsels Gallin, Michael Gottesman, the director of the intramural programs, and the NIH director on the Clinical Center. The board exercises oversight on the budget, resource allocation, planning, and operations. It also advocates for the Clinical Center within the NIH by emphasizing its unique purpose.

"The biggest lesion is that the intramural clinical investigators are not making use of the Clinical Center's unique resources," said Edward Benz, the chair of ABCR. "There's extraordinary high-tech imaging, the patients can stay as long as the investigators want, and the staff is dedicated to the patients and the research. However, the budget has been so flat in the past few years that the investigators now have many constraints on length of stay, and it has been hard to make capital investments, so that things like imaging equipment aren't as far ahead any more."[81]

The members of the ABCR are also responsible for recommending how career paths for clinical investigators in the intramural program can be improved.* To increase collaborations between intramural and extramural clinical investigators, the board is encouraging greater use of sabbatical leaves so that NIH investigators can work at other institutions, and extramurally funded investigators can work at the NIH.

The ABCR has 17 members: eight from the NIH, and nine from other institutions. Gallin, Gottesman, and the chairman of the medical executive committee serve as *ex officio* members. The whole committee meets for one day, three times a year, but working groups from the committee study Clinical Center activities at other times.† Gallin and his staff meet twice each month with all institute and center directors involved in Clinical Center matters, and conduct two retreats per year to discuss budget, the Roadmap, and the review process for grants.[57]

Education

The NIH Clinical Center offers, at no charge, courses in clinical research, clinical pharmacology, ethics, and institutional review boards.[57,84-86] The "Introduction to the Principles and Practice of Clinical Research" course may be taken at the NIH in 29 sessions over the course of four months or over the internet through a medical school approved by the NIH. In the most recent year, more than 900 students took the course, about 45 percent at the NIH and the rest over the internet.[86] Using archived lectures of the course, students at 12 international sites participated through agreements between their institutions and the NIH. Duke University School of Medicine offers the degree of Master of Health Sciences in Clinical Research in association with the NIH Clinical Center. This program uses live videoconferencing.

---

* As at universities, clinical investigators at the NIH frequently have more difficulty obtaining tenure than do basic scientists. Clinical research often takes longer to produce results that can be reported in the scientific literature and added to the candidate's CV.[81]

† The responsibilities of the members of the Advisory Board reach beyond the Clinical Center to the other institutes and outside the NIH to improve clinical research in the intramural program.[81,82]

John Gallin and Frederick Ognibene, the director of clinical research training and medical education in the Clinical Center, edit a text, now in its second edition, that forms the basis for the "Introduction to the Principles and Practice of Clinical Research" course.[86,87]*

The course in clinical pharmacology, which is offered yearly at the Clinical Center, runs one evening a week from September through April. Six hundred and thirty-five students took the course in 2006–2007, about 60 percent from the NIH and 40 percent through the internet.[86] A textbook titled *Principles of Clinical Pharmacology*, also in its second edition, forms the basis for this course.

The program will try to standardize among different institutions with the new clinical and translational science awards (CTSA)† how the results of clinical research are reported, how the institutional review boards operate, and how adverse events to patients are described.[88]

Department of Laboratory Medicine

The clinical laboratories in the Clinical Center perform about 1.8 million tests and 700 different assays per year, many of which are conducted in the support of clinical research. "We have unique opportunities," said the director Thomas Fleisher,‡ "because we're not pressed by economic pressures to concentrate on routine testing."[89] About 80 percent of the testing is standard, but 20 percent is performed for studies developed by investigators in the Clinical Center. When volumes fall below about one or two per day, the lab may outsource such work, often to the Mayo Clinical Laboratories, the diagnostic reference laboratory for the Mayo Clinic, which specializes in performing esoteric tests.

As the acuteness of patients' illness in the Clinical Center increased—a pattern established initially by studies focused on the therapy of lymphoid malignancies and the AIDS epidemic—the laboratory began operating 24 hours a day, seven days a week. Work was more leisurely in the 1950s and 1960s when much of the clinical research involved prolonged metabolic studies in which many patients spent weeks or months in the Clinical Center.[89]

---

* Academic Press, 2007.
† See Chapter 8.
‡ B.S., M.D., and pediatric resident, Minnesota; fellow in immunology, NIH.

Department of Transfusion Medicine

Although most hospitals have "blood blanks" where blood products are processed for transfusion into patients, few have departments of transfusion medicine where clinical research is strongly emphasized as well as clinical service. The director of this department at the NIH is Harvey Klein. He declared, "If not for the draft [Klein was a yellow beret] I likely would have remained at Hopkins," where he was a medical student, intern, resident, and fellow. He came to the NIH in 1973 for research training and stayed, deciding that "I liked NIH better."[70] As for the Clinical Center, where he now works, Klein said, "there's no other place like it." Despite being offered senior jobs at renowned medical schools, he has never left.[66] Klein calls his fellowship training program "Number one in academic blood transfusion."[70]

"Our primary role," said Klein, "is to provide services in blood and blood-support to the hospital. Our research is in areas related to the service we give."[66] Klein and his colleagues study the adverse effects that transfusions may produce in patients. They also prepare novel cells—for immunotherapy in transplant patients, for example. Since the beds in the Clinical Center are assigned to the institutes, not to support departments like transfusion medicine, when Klein is assigned to clinical duties, he sees patients of the National Heart, Lung, and Blood Institute, where he worked when he first came to the NIH.

Among the most striking results of work in this department was the discovery of the "Australian antigen" by Drs. Harvey Alter and Baruch Blumberg. Further work on the antigen led to its identification as the hepatitis B virus.* Alter later found another virus in the blood of patients who had developed hepatitis after transfusion. He named it "non-A, non-B hepatitis," now called hepatitis C, which became recognized as the leading cause of transfusion hepatitis.† Hepatitis C‡

---

* Which led to Blumberg's winning the Nobel Prize in Physiology or Medicine in 1976.
† Alter (B.A., M.D., Rochester, clinical associate at NIH, resident at University of Washington, fellow at Georgetown), chief of the infections diseases section in the laboratory of transfusion medicine, won the Albert Lasker Award for Clinical Medical Research in 2000 for this discovery.[90] Harvey Klein said of him, "As a young research fellow, Alter co-discovered the Australia antigen. For many investigators that would be the highlight of a career. For Alter it was only an auspicious beginning."[91]
‡ "Hep-C" to its devotees.

proved to cause cirrhosis in 20% to 30% of those infected over the course of 20 to 40 years and to evolve to liver cell cancer in some. End-stage chronic hepatitis C is now the leading indication for liver transplantation. Recurring infection in transplant recipients can be particularly virulent, with cirrhosis recurring in only three to five years. Alter speculates that suppression of the immune system, required in transplant patients so that they do not reject the transplanted organ, may cause the rapid progression of liver damage.[92]

Thanks to the work of Alter, who has spent most of his career at the NIH, and others, hepatitis now rarely develops from blood transfusions.[92] Alter explains why investigators at the NIH can obtain such results: "We can do prospective observational studies here whether or not they work out, so we could follow the transfusion story for decades."[92] Alter and his group are now studying a wide array of other transfusion-transmitted diseases to determine their incidence and means of prevention. "We're looking for any bug that can cause transfusion-transmitted disease," Alter explains, "and also looking for new transmissible agents whose disease outcomes may not yet be recognized."[92]

Despite this stellar research, the NIH receives few patients for liver transplantation, a procedure being conducted at many academic medical centers. Accordingly, Alter has collaborated with workers at nearby Georgetown University and more recently at the University of Pittsburgh, which have large liver transplantation programs. The transfusion program suffered when the NIH closed its cardiac surgery program, forcing the investigators to work with doctors at other hospitals.[92]

Surgery

Surgeon Steven Rosenberg,* a yellow beret from 1970 to 1972, who returned to the NIH in 1974 after finishing surgical training in Boston to become chief of surgery in the National Cancer Institute, has never left, an altogether unusual career path for an academic surgeon with his accomplishments. Over the years he has been recruited as chairman of surgery at some of the nation's leading medical schools where he would have earned far more money than the NIH can pay, but he has stayed in Bethesda because "I wanted to create tomorrow's medicine, and the

---

*M.D., Johns Hopkins; Ph.D. in biophysics, Harvard; surgical intern and resident, Peter Bent Brigham Hospital.

NIH was the place to do it."⁷⁴ Rosenberg has also turned down senior administrative positions at the NIH because "the next promotion means stopping science and no longer caring for patients."⁷⁴

"The NIH is absolutely unique for doing basic and clinical research," said Rosenberg. "We have no emergency department and no local community that depends on us for care as have most hospitals. We admit only those patients who can teach us something. The only reason we exist is to make progress."⁷⁴ Rosenberg works in his research laboratory almost every day and, when interviewed in the summer of 2007, had 15 patients with metastatic cancer on his service in the Clinical Center.⁷⁴

"Excluding skin cancers, benign tumors, and most non-invasive cancers, about half of all patients with potentially invasive cancers can be cured," Rosenberg explains. "Surgery will cure about 80 percent of them, but that's not much different from the 1970s. Chemotherapy and radiation help a bit."* With these data in mind, Rosenberg explored other means "where[by] progress might be made" and chose immunotherapy.

"The immune system is incredibly specific and sensitive, unlike chemotherapy, radiation, and surgery, which kill normal as well as cancer tissues." However, immunotherapy for cancer is "in the infancy of its development," Rosenberg admits, despite all the years and effort that he and others have given to it. The method, however, does appear promising in the treatment of melanoma, a universally fatal skin disease when it becomes metastatic.⁷⁴,⁹³†

While Rosenberg has stayed at the NIH, other surgeons have left. Neurosurgeon Edward Oldfield departed after 26 years at the NIH, 21 of them as chief of the clinical neurosurgery section and of the surgical neurology branch. Oldfield stayed there for as long as he did because, "It's a great place. One can focus one's time on the science one wants to

---

*Vincent DeVita, who as National Cancer Institute director in the 1980s was Rosenberg's administrative chief, said, "I used to argue with Steve that his statements about cancer cures were too negative. The overall survival rate increased from 35% to 65% in this period, and the national mortality rate has fallen."²⁸ DeVita cites impressive improvement in the treatment of colorectal and breast cancer—"we do many fewer colostomies and mastectomies than before"—and for Hodgkin's disease and some of the leukemias, and even for lung cancer, where, although the results remain depressing, survival has increased from 6% to 12%.²⁸

†Rosenberg has described his work in the book *The Transformed Cell. Unlocking the Mysteries of Cancer*, New York, G. P. Putnam's Sons, 1992.

do."[94] He left more for personal than for professional reasons. When he was invited to become professor of neurosurgery and internal medicine at the University of Virginia, he accepted.

"The NIH is seen as a molecular institution," said Oldfield. "People begin to think that clinicians who work at the NIH for a long time become basic scientists, and this reduces their attractiveness to clinical departments at medical schools."[94] When he left, the NIH was paying Oldfield about $250,000 annually, plus between $20,000 and $30,000 as an annual bonus, substantially less than he could make as a professor at a medical school or in private practice.

Investigating Fibrous Dysplasia of Bone

The work of Michael Collins is typical of an NIH investigator in the Clinical Center.* Trained first as a musician and then as a physician, Collins first worked at the NIH as a biologist before going to medical school. He returned for his fellowship in endocrinology and metabolism in 1995 and then joined the craniofacial and skeletal diseases branch of the National Institute of Dental and Craniofacial Research as a clinical investigator. He is currently a tenure-track investigator and chief of the skeletal clinical studies unit in the dental institute.

Like many NIH researchers in the Clinical Center, Collins studies a rare disease. His is called "fibrous dysplasia of bone," in which one or more bones are weakened by replacement of normal bony tissue with fibrous tissue. The lesion causes the affected bones—only one bone in most patients—to expand and deteriorate, often leading to painful fractures. When the disease involves the skull or facial bones, obvious deformities can appear. Depending upon the severity of the disease, it may first be recognized when the patient is a child, adolescent, or adult. Some patients are so disabled that they arrive at the NIH in wheelchairs. Others first learn that they have the disease from an X-ray performed for an unrelated injury such as an automobile accident or from one of the endocrine disorders with which it may be associated, such as precocious puberty.

The genetic basis of fibrous dysplasia is known, but the disease is not inherited. There is no specific treatment as yet for the bone disease,

---

*Disclosure: I first learned to admire the skills of Collins when he was a medical student and trainee at the University of Maryland.

although orthopedic surgery can relieve symptoms and strengthen abnormal bone in some patients. In most cases, one can treat the endocrine disease that many have with medicine, surgery, or a combination of the two.

Eighty percent of the patients who come to the NIH with this disease refer themselves after someone has studied the internet. Doctors send about 20 percent of them. The patients are usually admitted on Monday and stay about a week. On average, two or three are being investigated in the Clinical Center at one time.

Collins personally—often accompanied by a fellow—takes a history, examines the patients, and studies the data that doctors caring for the patients have sent him. The patients then begin several days of blood and imaging tests and a series of collaborative consultations. The study's research nurse has scheduled each of these in advance and has also made sure that pre-admission information on each patient has been received before admission.

A social worker has documented the financial resources of the patient and family to determine how much assistance is required to bring the patient to the NIH. "Once they get here," said Collins, "we take care of everything."[40]

"We're seldom fooled about the diagnosis," he declared, both because so much has been learned about every patient before admission and because he has had so much experience with this unusual disease. "Sometimes, we recognize 'look-alikes,'" patients with similar findings but different diseases.[40] Collins will refer them to investigators at the NIH or at other institutions with the appropriate expertise.

Toward the end of the week the team—Collins, the fellows, and the nurses—reviews the data to be sure they have all the facts concerning the patient that might reveal useful information about the particular patient and about the disease itself. In many cases, a surgical operation, a trial of an investigative drug, or an "off-label use"* of an existing medicine may be advised. Each of the patients knows, however, that they have contributed to increasing the understanding of their disease, and each has had the superb care for which the Clinical Center is widely recognized.

---

*A drug used "off-label" is given to treat an illness other than those for which the Food and Drug Administration has approved its use.

Each week, members of the team review their experience with the patients studied that week. Data are available for more than 20 years on some patients, giving the NIH investigators an unrivaled opportunity to study the natural history of the disorder. When effective treatments, some of which are under development at the NIH, become available, the team will be able to compare how the patients have done before and after that happy day.

Like so many investigators at the NIH, Collins can't imagine a better place to work. "I can go from idea to clinical study in sometimes as little as a few months," Collins said. "If need be, I can keep patients in the hospital for weeks for detailed evaluations. This would be very difficult to do anywhere else."[95]

Ethics

Each of the institutes and centers has a deputy ethics counselor, ("DEC" in NIH-ese) who reports to her or his director for ethics issues. The DECs train each employee in appropriate ethical behavior and are available to advise employees when ethical questions arise. They also review the relevant financial holdings of all investigators involved with clinical trials. The Department of Health and Human Services approves the appointment of each DEC.[61]

Although the NIH has had a bioethicist program in the office of the director since 1977, the Department of Clinical Bioethics, created in 1995, resides within the Clinical Center.[96,97] Its chief is Ezekiel Emanuel,* who was associate professor of social medicine at the Harvard Medical School before he joined the NIH in 1998.

Emanuel is not shy or retiring when describing the quality of his unit. "We've got the best department in the world, but we're fragile since we can't pay competitive salaries to some people we'd like to recruit. I can compete successfully for professors in philosophy departments but never for a tenured professor at a law school."[98] Nevertheless, the security of an ethics job at the NIH, relief from teaching, great freedom to work how one wants, and a collegial atmosphere have enabled Emanuel to fill his unit with highly competent people. "No dummies here," he said."[98]

---

* M.D. and Ph.D. (political philosophy) Harvard; M.Sc. (biochemistry) Oxford (Exeter College); training in internal medicine and oncology, Beth Israel Hospital and Dana-Farber Cancer Institute, Boston; fellow in ethics and the professions, Kennedy School of Government, Harvard.

Chapter Four: The Centers   133

National Center for Research Resources (NCRR)

Established in 1962

2008 budget (in millions): $1,155

Intramural program: None

The National Center for Research Resources (NCRR) began as the Division of Research Facilities and Resources (DRFR) on June 15, 1962 and was renamed the Division of Research Resources (DRR) in 1969. In 1990, DRR combined with another unit, the Division of Research Services, to form the National Center for Research Resources (NCRR). The director (2008) is Barbara Alving.*

NCRR, a noncategorical center, provides extramural support for the infrastructure that research institutions need to conduct biomedical research.[41] Its work is concentrated in four areas: biomedical technology, clinical research, comparative medicine, and research infrastructure. It supports the purchase of large, expensive instruments like electron microscopes and mass spectrometers.† Whereas NCRR reviews internally many of the grants submitted to it, the Center for Scientific Review evaluates most of its investigator-initiated R01 grants as well as applications for instrumentation and biomedical technology resource centers.[41] NCRR pays for maintaining eight primate research centers at different locations throughout the country.

Twenty-three states and Puerto Rico, which have relatively little extramural support, receive Institutional Development Awards from NCRR. Most of the recipients, like Alaska, Montana, the Dakotas, and Maine, are predominately rural and sparsely populated. The program, designed "to enhance the caliber of scientific faculty at research institutions and undergraduate schools, thereby attracting more promising students to these organizations,"[99] serves to calm the congressmen who feel that their states are underrepresented in the support they receive from the NIH. NCRR also supports research centers in predominately minority medical schools such as Morehouse, Meharry, Drew, and the schools in Puerto Rico.

---

* M.D. and medical intern, Georgetown; medical resident and hematology fellow, Johns Hopkins.
† A spectrometer is used to determine the composition of a physical sample.

NCRR also sponsors Science Education Partnership Awards (SEPA) which are designed to improve life-science literacy among senior high school students and teachers. Science centers and museums can also receive funds in this program.

NCRR administers the new Clinical and Translational Science Awards and the older General Clinical Research Centers that are being phased out.*

Center for Information Technology (CIT)

Established in 1964

2008 budget (in millions, taxed from institutes): $218

Intramural Program: three senior investigators

The Center for Information Technology, which was established as the Division of Computer Research and Technology (DCRT) in 1964 and assumed its current name in 1998, provides, coordinates, and manages information technology at the NIH.[100] CIT does not have an extramural program.

The center has a relatively small intramural program of about 60 statisticians, engineers, biologists, and staff to assist members of the intramural program in their computational needs.[101] "This was a legacy from [one of the former units] that became part of CIT," said Alan Graeff, the director from 1998 to 2005. "The program was created when most research scientists didn't have personal computers, not to mention clusters and other forms of higher-performance environments that have now become part of the fabric of modern research laboratories."[101]

Graeff became the third permanent director of CIT and its predecessors and the first chief information officer at the NIH.[102] Graeff, who had started at the NIH as a biologist in 1973, began working in computer technology in the 1980s. This led to his becoming deputy chief of the information technology branch of the National Institute of Allergy and Infectious Diseases in 1987 and chief of the information systems branch at the Clinical Center in 1995 before directing CIT. In

---

* See under Roadmap, Chapter 8.

2005, Graeff left CIT to become deputy director of the National Center for Biotechnology Information, in the National Library of Medicine (NLM).*

In the winter of 2008, the acting director of CIT was John "Jack" Jones.[103]

John E. Fogarty International Center (FIC)

Established in 1968
2008 budget (in millions): $67

Intramural program: None

John Fogarty, a master bricklayer and president of his local union, was elected to Congress from the second district of Rhode Island in 1939 at the age of 26 and served for 27 years. As chairman of the House Appropriations Subcommittee for Labor, Health, Education, and Welfare, Fogarty advocated strongly for the NIH and for medical research in general, with particular emphasis on global health. After he died suddenly at his desk in January of 1967 at the age of 53, his colleagues authorized funding for the John E. Fogarty International Center for Advanced Study in the Health Sciences at the NIH.[104]

Initially, the Fogarty Center funded grants offering stipends and travel and living expenses to support senior and mid-career scholars, usually from countries other than the United States, to come to Bethesda and participate in the activities of the NIH. The Scholars-in-Residence program brought 200 eminent scientists to the NIH, including Margaret Mead, the cultural anthropologist; Sir George Pickering, Regius Professor of Medicine at Oxford University: Albert Sabin, who developed a vaccine for polio; four Nobel Prize winners; and several Lasker winners.

The first group lived together in Stone House,† a large Colonial Revival mansion built in 1930 for the Reverend George Freeland Peter, a canon of the Washington Cathedral, whose family had

---

* See http://www.ncbi.nlm.nih.gov/ for more information about the National Center for Biotechnology Information.
† So called for the locally quarried bluestone on the exterior of the building.

owned the property since the eighteenth century. The government, which acquired the estate in 1949 to provide space for an expanding NIH, named it, in 1989, the Lawton Chiles International House to honor the senator from Florida (1971–1989), who had been a staunch supporter of the NIH. Originally a department in the director's office, the Fogarty became a center in 1984–1985.

From 1970 to 1978, the scholars lived on the second floor of Stone House and ate meals together in its dining room. When the program expanded, the building ceased to serve as a residence, and the second and third floors now have offices for some of the Fogarty staff.

The Fogarty's current mission is much broader than the original Scholars-in-Residence program, which continues for a few distinguished investigators each year and is now supported entirely by the institutes and centers.[105,106] With the arrival of the HIV/AIDS epidemic, the Fogarty took on the role of building research capacity in low- and middle-income countries. The center supports programs in tuberculosis, malaria, HIV/AIDS, and other infectious diseases, and funds research training for scientists from countries in which it has programs.[105]*

The director of the FIC since 2006 is Roger Glass,† an international infectious disease specialist and an authority on rotaviruses and noroviruses, important causes of diarrheal diseases in underdeveloped as well as developed countries. Glass has conducted field studies in India, Bangladesh, Brazil, Mexico, Israel, Russia, Vietnam and China. He is fluent, and often lectures, in five languages. Glass is currently conducting a strategic planning process for the center, which will expand its activities in the future.[104,107]

National Center on Minority Health and Health Disparities (NCMHD)

Established in 2000

2008 budget (in millions): $201

No intramural program

---

* For details about the center's many programs, see the Fogarty Center's website http://www.fic.nih.gov/.
† M.D., M.P.H., Harvard; Ph.D., Goteborg, Sweden.

Congress established the Office of Research on Minority Health at the NIH in 1993 and advanced it to the status of a Center seven years later. John Ruffin, its first director, was appointed in 2001.

As stated in its website:[108] "The National Center on Minority Health and Health Disparities (NCMHD) promotes minority health and leads, coordinates, supports, and assesses the NIH effort to reduce and ultimately eliminate health disparities. The NCMHD works independently and in partnership with the NIH Institutes and Centers and with other Federal agencies and grassroots organizations in minority and in other medically underserved communities to: conduct and support basic, clinical, social, and behavioral health disparities research; promote infrastructure development and training; foster emerging programs; disseminate information; and reach out to minority and other health disparities communities."

National Center for Complementary and Alternative Medicine (NCCAM)

Established in 1999

2008 budget (in millions): $122

Intramural program: one tenured senior investigator

Senator Tom Harkin, who has a strong interest in dietary supplements, and former congressional representative Berkley Bedell (D-Iowa) were the moving forces in Congress to obtain funding to establish an office within the NIH to investigate and evaluate complementary and alternative medicine.[109]* Many NIH leaders opposed the creation of such a group,[112,113] believing that a program in alternative medicine was unsuitable for their institute; "that the advocates just wanted the imprimatur of the NIH on their pet projects"; and that "some of them were more interested in politics than science."[110]†

---

* Senator Harkin has been, over the years, one of Congress's most ardent supporters of NIH projects and larger budgets. His influence in bringing about the creation of new institutes, centers, and offices has led one observer to remark that he "tried to run the NIH from the Hill."[110] Harkin and Senator Arlen Specter continue to support NCCAM strongly.[111]

† Opposition to NCCAM continues. See I.V. Atwood, KC. "The ongoing problem with the National Center for Complementary and Alternative Medicine." http://csicop.org/si/2003-09/alternative-medicine.html.

The use of such preparations is widespread, however. A 1993 report from the Harvard Medical School showed that 34 percent of a sample of 1,539 adults, "reported using at least one unconventional therapy in the last year" and that "72 percent of the respondents who used unconventional therapy did not inform their medical doctor that they had done so."[114] More recent data show that consumers spend more that $23 billion each year "on natural products marketed to maintain or enhance health."[115]

Since many of the medicines that the advocates champion were directed toward the treatment of cancer, they favored attaching the new entity to the National Cancer Institute. Instead, the NIH assigned the entity to the Director's office and gave it center status in 1999.

Research about alternative medicine continues to be conducted in some of the institutes, however. The National Cancer Institute, for example, has not transferred its studies to NCCAM, and it directly supports extramural trials on the effectiveness of alternative and complementary products. The NCI spent about $120 million for this work in 2008, mostly in the extramural program. As of the spring of 2008, the Office of Cancer Complementary and Alternative Medicine in the NCI was not supporting clinical trials in the Clinical Center through the intramural program.[116]*

Directors

Joseph Jacobs was appointed the first full-time director of the Office of Alternative Medicine (OAM) in 1992. Jacobs,† an American Indian, was familiar with indigenous medicine and had designed clinical trials. He had, however, no specific experience with alternative medicine. The budget was small, only $2 million in the first year.[110]

Jacobs, tired of the political battles that the office and the advocates spawned, served only two years. "I didn't think they should be a separate group," Jacobs still believes. "The office was 'ghettoizing' the topic, which should be mainstream in the existing institutes. This

---

* See the website for the Office of Cancer Complementary and Alternative Medicine for more information about the program: http://www.cancer.gov/cam/cam_annual_report.html.

† M.D., Yale; pediatrics resident, Dartmouth and Yale; Robert Wood Johnson Foundation Clinical Scholars Program, University of Pennsylvania. Jacobs is now associate medical director at Abbott Molecular, Des Plaines, Illinois.

would have been more efficient and would have produced better, unbiased science."[110]

Stephen Straus was the founding director of NCCAM when the office was elevated to the status of a center in 1999. Straus, chosen after a long search, "took a controversial office and created something good," said Josephine ("Josie") Briggs, the current director.[111]

When Straus resigned in 2006,* Ruth Kirschstein became acting director, pending the recruitment of Briggs, a nephrologist, as the second director two years later. Briggs was returning to the NIH, where she had been director of the division of kidney, urologic and hematologic diseases in NIDDK from 1997 to 2006. For the next two years, she worked at the nearby Howard Hughes Medical Institute. When asked why she decided to lead this most controversial of the institutes and centers, Briggs laughed and said,

> Perhaps it reflects the fact that I am a child of the '60s. I am interested in ideas that come from outside the mainstream, and I believe NCCAM's mission is a valuable one for the American public.[111] ... Despite the skepticism of most doctors that these treatments are efficacious, Congress and the public continue to support our work since these approaches are important to so many Americans. Building evidence is what we're about. We must bring rigorous science to these issues. Although we are not a regulatory agency, we do provide information to help consumers make informed health care decisions.[117]

Purpose and Topics

"NCCAM's purpose is to build an objective evidence base concerning complementary and alternative medicine," said John "Jack" Killen, the acting deputy director of NCCAM and director of the center's Office of International Health Research. "We're open to the possibilities that health benefits can come from many different roots, but require scientific validation of safety and usefulness."[118]†

---

* Straus died of a brain tumor not long afterwards.
† See the Institute of Medicine report of 2005, "Complementary and alternative medicine in the United States," http://www.iom.edu/CMS/3793/4829/24487.aspx.

"And we must tell the public about those remedies that don't work or are dangerous," Josie Briggs added. "We're the people to do this."[111] On the center's website* are reports of studies on the efficacy and mechanism of action of several nontraditional remedies. Among these are:

- **St. John's wort.** Two large studies, one sponsored by NCCAM, showed that the herb was no more effective than a placebo in treating major depression of moderate severity. NCCAM is studying the use of St. John's wort in a wider spectrum of mood disorders, including minor depression.
- **Acupuncture.** Despite many studies on acupuncture's potential usefulness, results have been mixed because of complexities of study design and size, as well as difficulties with choosing and using placebos or sham acupuncture, according to the NIH Consensus Statement on Acupuncture. An NCCAM-funded study recently showed that acupuncture provides pain relief, improves function for people with osteoarthritis of the knee, and serves as an effective complement to standard care.
- **Glucosamine plus chondroitin sulfate.** A study published with the support of NCCAM in 2006 showed that glucosamine plus chondroitin sulfate did not provide significant relief from pain due to osteoarthritis among all participants. However, a smaller subgroup of study participants with moderate-to-severe pain showed significant relief with the combined supplements.
- **Spinal manipulation for lower back pain.** Overall, studies have shown that spinal manipulation can provide mild-to-moderate relief from lower back pain and appears to be as effective as conventional medical treatments.†
- **Asian ginseng.** To date, research results on Asian ginseng are not conclusive enough to support health claims associated with the herb. Some studies have shown that Asian ginseng may lower blood glucose. Other studies suggest possible beneficial effects on immune function.

---

* NCCAM's website is http://nccam.nih.gov/. Catherine Law, NCCAM science writer and press team leader, provided additional information.[119]

† See http://nccam.nih.gov/health/pain/spinemanipulation.htm#science for additional information from NCCAM about spinal manipulation for lower back pain.

- **Ginkgo.** Some promising results have been seen with the use of ginkgo for Alzheimer's disease/dementia, intermittent claudication, and tinnitus, among other problems, but larger, well-designed research studies are needed.
- **Echinacea.** Studies have shown that echinacea may be beneficial in treating upper respiratory infections, but echinacea does not appear to prevent colds or other infections.
- **Black cohosh**, whether used alone or with other botanicals, failed to relieve hot flashes and night sweats in postmenopausal women or those approaching menopause. The results from other studies vary in showing whether black cohosh effectively relieves menopausal symptoms. There are not enough reliable data to determine whether it is effective for rheumatism or other uses.

CHAPTER FIVE

Finances

THE NIH IS an agency of the Department of Health and Human Services, which, except for the Department of Defense, has the largest federal budget of any government entity. In fiscal year 2008, the NIH received $29.464 billion, about four percent of DHHS's budget.

The NIH has only one source of money, the federal government.[1] Consequently, NIH workers cannot receive support from sources outside the NIH, such as foundations and industry, as can investigators at universities. NIH investigators, however, can ask institutes other than their own for additional resources,[2] and the few whose discoveries have become commercially successful are permitted to earn royalties from industry.

The White House develops the budget, but Congress often modifies it, usually upward (at least in the past).[3] President Clinton told Harold Varmus that he always "underplayed it because Congress will increase it."[4] Congress loves the NIH, while presidents try to rein it in.[5]

At their meetings with the appropriations committees, the institute directors loyally present what the current administration has prescribed. Later, congressmen often ask for the scientists' "professional" judgment of the NIH's needs. This gives them the opportunity to suggest that larger amounts could be usefully employed.[6]

As an agency of the Department of Health and Human Services, the National Institutes of Health and the NIH director report to the

Secretary of the department, and through the Secretary to the President. Unlike boards of directors in the private sector, the NIH advisory boards have no fiduciary responsibility.

Before 1995, the agencies of the Public Health Service, which included the NIH, reported to the Assistant Secretary for Health—Philip Lee from the University of California–San Francisco from 1993 to 1997—who reported to the Secretary.[7,8] Seeing the Assistant Secretary's office as an administrative obstruction between him and the Secretary of Health and Human Services, Varmus favored its elimination. Kenneth Shine, then president of the Institute of Medicine, disagreed. "The lack of this representation 'downtown' [in the Department of Health and Human Services] became an increasing problem in future administrations. Harold Varmus had a unique relationship to the Secretary that could not be replicated subsequently."[8] Shine said about Lee, "No one in DHHS knew science and medicine as well as Phil."[8]

The Doubling

From 1998 to 2003, the NIH enjoyed five years of extraordinary financial growth as Congress allocated enough money to double the budget from $13.648 billion to $27.067 billion. "It came to pass," one of the institute directors said, because of "an alignment of stars," and he credits President Clinton, Department of Health and Human Services Secretary Donna Shalala, NIH Director Harold Varmus, congressmen John Porter and Newt Gingrich, and senators Tom Harkin and Arlen Specter for making it happen.[9] Another important congressional supporter was Senator Connie Mack (R-Florida, 1989–2001) who introduced, in 1997, the Biomedical Research Commitment Resolution, which called for doubling the budget in five years.[10]* "And don't forget the advocates who believe that if you put money into research, you get cures," adds science journalist Barbara Culliton.[11]

The increase of about 15 percent per year substantially exceeded the average growth of about eight and a half percent per year from 1971 to 1997. To double the NIH budget had taken about twice as many years before the 1998 to 2003 bonanza.[12] Understandably, the doubling

---

*Personal experience with disease led Mack, like many advocates, to support the NIH. His father died of esophageal cancer, and Mack's wife was a breast cancer survivor. His brother had died of melanoma, for which Mack was treated in 1989.[10]

wasn't received with enthusiasm at other agencies, like the Centers for Disease Control and Prevention (CDC), which didn't participate in the largess.[13]

After the September 11, 2001, attacks, significant amounts of the doubling were assigned for research on biodefense and terrorism.[9,14,15] Since some of the money was awarded to companies and non-NIH agencies, NIH investigators saw these funds as "laundered" through the NIH budget for assignment elsewhere. Some of this money, however, stayed within the NIH, mostly in the National Institute of Allergy and Infectious Diseases.[14] Since the funds were not available for typical NIH research,[15] some objected to the assignment of the money to biodefense.[16]

A soft landing with traditional growth of six to eight percent after the doubling did not occur, and, some have suggested, shouldn't have been expected.[11,17] Since fiscal 2003, the last year that the doubling applied, the budget, once inflation is included, has shrunk each year.[18] In fiscal 2008, it was $29.464 billion, an increase of $2,389,305 since the doubling stopped, an average of $478 million per year. Since on an annual basis this is only 1.8 percent before inflation is factored in, the NIH and its institute directors have relatively few uncommitted funds.

In recent decades, the NIH has never had to endure a similarly lengthy period of declining purchasing power in its budget.[18,19] The year 2007 brought some relief, with the budget increasing by 2.4 percent to $29.236 billion, still less than inflation; but in 2008, the increase was only 0.8 percent. John Porter, a strong NIH supporter while a congressman and afterwards, would like to see funding for the NIH return at least to its "historical rate of increase of three percent plus inflation."[20]

Even though the doubling had favored the extramural more than the intramural program,[9,21] investigators and institutions supported by the extramural program and the congressmen who represented them complained bitterly when the flat budgets appeared.[21,22] These events have made some wonder whether the doubling was wise.[11,19,23-32] Said some observers:

- "Doubling was a revved-up version of the old historic pattern. We warned that a hard landing could ruin it all. Maintaining the old pattern would have been better."
- "Better to have more predictable growth than doubling, then flat."
- "Doubling gave a false feeling that dollars would always come."

- "The doubling occurred too fast. It backfired. Sustained growth of eight to nine percent would have been better."
- "We should have spent the money in new areas rather than just adding dollars to what was running."
- "The doubling was a bubble."
- "There's tragedy in getting what you want as well as not getting what you want."

By 2008, the NIH would have received the same amount of money by growing ten percent each year as by doubling and then enduring the flat budgets that followed.[12,33] Of course, this presumes that the budgets would have grown at the historic rate during each year from 1998 to 2008. Response to the attacks of September 11, 2001, the wars in Afghanistan and Iraq, tax reductions, deficits, and the priorities of a Republican administration and Congress might have frozen NIH budgets anyway, without the benefit of the doubling from 1998 to 2003.

One observer said, "It is not at all clear that [without the doubling] NIH would be getting larger increases now."[33] But how could one have predicted this when the doubling was being enacted with great enthusiasm in the 1990s?

Doubling wasn't the entire problem. It was what followed that caused the heartache.[15,17,34,35] *The New York Times* editorialized in March, 2008, "Neither the government nor academia gave much thought to what might happen when the flush times came to an end."[36] "Not true at the NIH," responds Michael Gottesman. "We had extensive discussions about how to create a soft landing, but given annual budgets that had to be spent, this proved to be very difficult."[37]

Since money that the NIH does not spend by the end of the fiscal year will be lost to the NIH, the directors and investigators try to spend all budgeted funds before then. The funds cannot be "saved" and spent in the next year.

A Straitened NIH*

With the NIH profiting from the doubling and then struggling with flat budgets—actually decreasing available funds when inflation is taken

---

* This book ends with the 2008 calendar year before the Obama administration and its friends in Congress substantially increased the NIH budget.

into account—critics of the process complained that the leadership did not develop enough explicit plans for new programs other than increasing the number and size of investigator-initiated (R01) grants.[31,38] Complicating flat budgets has been the policy by which the NIH pledges funds for several years to successful applicants of the extramural program. Since these grants must be supported despite the size of the budgets, funds available for new applications, which have consistently become more expensive in successive years, drops.[39]

"The doubling was essential but mismanaged in not planning for a soft landing," says Kern Wildenthal, president of the University of Texas Southwestern Medical Center at Dallas.[40] Others agree.[24,32] "We deluded ourselves with the doubling, thinking that such annual increases would go on forever, at the very least protecting researchers from the cost of inflation," said Arthur Levine, scientific director of the National Institute of Child Health and Human Development from 1982 to 1998.[41]

Advocates and members of Congress are calling the NIH to account for not using the additional funds to find specific cures for the diseases in which they have an interest.[38] Most of the funds in the doubling were directed toward supporting investigator-initiated research in the extramural program. Scientists know that such work seldom leads quickly to the cures that Congress and advocates hoped the money would produce.[42]

Leading biomedical scientists complained vigorously about the ill effects of the government's reduction in support for research.[24,43] "There's a feeling of insecurity among scientists," one of the institute directors said.[44] Nobel Prize winner Michael Brown believes that "morale among scientists is very low. Though it goes in cycles, it's worse now."[24] NIH leaders and advocates regret that no one in the House of Representatives has replaced John Porter, who retired from Congress in 2001, to champion the NIH as he did.[23] Some congressmen and their aides see the NIH as just another lobbying group with an insatiable need for money.[32]

Part of the angst at the medical schools was self-induced. With the doubling of the NIH budget came more faculty, trainees, and buildings.[11,18,23,45] As one former institute director says, "what deans most like to see are [construction] cranes on their campuses."[46] David Korn adds, "The strategy in academic medicine has been growth, all predicated on the NIH growing. All that capacity had been built. Now the success rates are way down, and times are tough."[19]

As the doubling ended, the debt service on the new laboratories came due, from about $3.5 million per research-intensive institution in 2003 to a projected $7 million per institution in 2008, according to a study by the Association of American Medical Colleges.[18] Just as new faculty and trainees, recruited as the money was flowing, began applying for support, flat budgets appeared.[3] Not having developed business plans to deal with the potential ending of the doubling,[11,31,40] the schools now had more trained investigators who were unable to win the grants that they needed in new buildings requiring indirect costs to help pay the interest and borrowed principal.

Graduate students and postdoctoral fellows see their mentors continuously writing grants[8,47] and wonder whether there is a career for them in academia[8,18,48,49] "at an unprecedented time of opportunity for biomedical discovery."[18]

- "Our investigators spend half their time writing or rewriting grants these days," said Philip Pizzo, dean of the Stanford University School of Medicine.[50]
- "Some of my colleagues are spending as much as 60% to 80% of their time writing grants," declared Joan Siefert Brugge, chair of the department of cell biology at Harvard. "The NIH administrators are not aware of this situation. Elias [Zerhouni] said that senior investigators have many grants and move the dollars around. It's not so."[51]
- "The best people don't want to serve" on study sections, said one former NIH investigator. "They're home writing grants."[41]

Even at the NIH, some institutes became overextended during the doubling.[21] Although institutes were able to acquire some equipment,[52] long-standing salary constraints caused significant unhappiness.[22,52] To address this issue and improve its ability to recruit, the NIH has been able to increase salaries recently.[53]

Until the financial climate improves, the NIH is:

- Canceling the inflationary increases traditionally added to the budgets of currently active grants, an annual process known in NIH jargon as "a noncompeting renewal."
- The savings will be directed toward funding as many new awards as possible.[54]
- Decreasing by ten percent the budgets of many of the National Cancer Institute's cooperative clinical study groups, thereby reducing by 3,000 the number of patients to be entered in studies.[55]

- Holding or reducing funding for 16 clinical trials of children with cancer.[55]

Consolidation and Outsourcing

Early in the administration of Tommy Thompson as Secretary of DHHS in the George W. Bush administration, the department challenged its agencies to become more efficient and save money by consolidating or outsourcing services. Although the authority to do this had long been the prerogative of the Office of Management and Budget (OMB) under a circular known as A–76,[56] different administrations had applied it with varying degrees of enthusiasm. Now the agencies were assigned annual quotas of savings.

To carry out the program, the NIH had to create a new office to administer the administration's policy.[57] This effort, which has been a recurring issue in several administrations, was designed to gain economies of scale. As acting NIH director at the time, Ruth Kirschstein was powerless to stop the program, although she did limit its effect.[58]

Of all the consolidations, human resources (HR) was the most "gut-wrenching" for the NIH, said Colleen Barros, the NIH chief financial officer and deputy director for management.[59] Before consolidation, each institute and center had its own HR officers. Now there is one HR office for the entire NIH.

"I used to have my own HR person," complained one of the institute administrators. "Now everything's slower since you have to get more approvals. It's less personal." One of the institute directors said, "When I arrived we had five full-time HR people. Now none, so there's no person on site to help our people. This means that more administrative stuff is falling on the investigators. Central HR doesn't understand what scientists do."[44] As a result of the consolidation the HR function was significantly impaired but is beginning to regain its previous level of function.

Another governmental project that affected the NIH was the effort of the Bush administration to order various agencies to compare the costs of operating certain functions internally versus by companies in the private sector. "It was done in the name of efficiency, less government, more private sector," said Norka Ruiz Bravo, deputy director for extramural research.[60]

One of the activities selected for this process was the extramural administrative staff in the institutes and centers[57,59,60] whose more than

Chapter Five: Finances

900 people support the extramural scientists and the directors who administer the NIH extramural program. Now the NIH had to make a formal bid to continue this work as if it were a contractor.

The NIH won the competition. The Division of Extramural Activities Support (DEAS) was created to centralize secretarial pools, clerks, and assistants. "Initially it was awful," said Ruiz Bravo. "People were very angry, and there was much conflict."[60] Because of the uncertainties with the process and implementation of the new division, some of the best employees retired or transferred to other jobs in the NIH or other government agencies.[57]

Under the new scheme, the institutes and centers can no longer directly recruit their staffs; this is done by DEAS, although the institutes can choose from among those whom DEAS proposes.[57] Six hundred and seventy-five are now doing the work that 900 did before the consolidation. "Despite the agony, it *has* decreased costs," said Ruiz Bravo;[60] but "it perturbed the NIH profoundly particularly because it touched science," added Barros.[59] One custom that the new system curtailed was the tendency of scientists in the institutes and centers to hire an additional employee to compensate for unsatisfactory performance by the person currently in the position, rather than discharging and replacing the employee.[57]

Many at the NIH objected to and have not as yet accepted as wise the government's administrative changes. As one NIH administrator* said:

> The current [George W. Bush] administration has as its mantra that downsizing of government must happen and is a good thing, regardless of the results. [Senior officers at the NIH] are mandated to report that the program has been fully successful, and are keeping to the party line on this issue. All of our good staff have been eliminated. We are left with a centralized, unsupervised low-level clerical staff. We no longer have a higher level of support staff to help with our jobs. All of the ICs [institutes and centers] have to find somewhat devious ways to work around the senseless administration mandates of this type, for example by hiring outside contractors, which is exactly what the administration wants. The program is largely a failure from the NIH staff point of view.

---

* Who, understandably, did not want to be identified.

The angst notwithstanding, the NIH estimates that consolidating has saved the government about $60 million, which includes the cost of administering the program, during the three-year period beginning in October, 2004.[57]

Technology Transfer

Although a government agency, the NIH can grant licenses to companies to commercialize technologies invented in the laboratories of its intramural program. In addition, its employees can collaborate with private companies in the research and development of medical products.[14,61] By 2007, almost 200 products that had reached the market included discoveries made at the NIH.[62]

Among the innovations developed in NIH laboratories that have been commercialized in the public market are:

- The first vaccine against the human papilloma virus (HPV), a major cause of cervical cancer.
- The first vaccine against the hepatitis A virus.
- A multidrug resistant (MDR) tumor line used to test chemotherapeutic drugs for their ability to overcome MDR.
- Therapy for multiple myeloma.
- Treatment for retinitis (inflammation of the retina of the eye) due to cytomegalovirus.
- Treatment for non-Hodgkin's lymphoma by radioimmunotherapy.

The process whereby NIH discoveries are transferred to the private sector is carried out in the Office of Technology Transfer (OTT), which is a branch of the director's office. OTT has about 70 employees.

Some of the institutes and centers also have technology transfer offices with one to 50 full-time or part-time employees per institute. The largest of these offices are in the National Cancer Institute; the National Heart, Lung, and Blood Institute; the National Institute of Allergy and Infectious Diseases; and the National Institute of Diabetes and Digestive and Kidney Diseases. The offices in the institutes receive the original reports of inventions and forward them to the central office for review and final decisions about whether to file a patent. The institute offices have the primary responsibility to develop collaborative agreements with industry.

"The picture most people get wrong about technology transfer is that we develop drugs and then give them to companies that make money," said Mark Rohrbaugh, the director of the Office of Technology Transfer. "What the NIH investigator provides is the basic concept that might, after much time and investment by the company, possibly result in a commercially successful product."[62]

A Cooperative Research and Development Agreement (CRADA) is often the vehicle used to facilitate the collaboration between the NIH and the private sector for transfer of federal technology to the marketplace.[63] A CRADA links one or more laboratories in the Public Health Service, which includes the NIH and the Food and Drug Administration, with an extra-governmental party. It allows the NIH to receive money from external sources that are developing a product. Without a CRADA, the NIH cannot accept the funds.

After the NIH has filed a patent application for the discovery, the Office of Technology Transfer then looks for a company to develop the invention. About 50 percent of these companies are small businesses.[62]

The Foundation for the NIH, a not-for-profit entity established by Congress in 1990, helps the NIH develop partnerships with private corporations.[64–66] The foundation has been "key to the development of several public-private programs during my tenure," said Elias Zerhouni. "For example, the Gates Foundation–FNIH Grand Challenges in Global Health, which directed 450 million dollars to this research, as well as several industry–NIH collaborations."[39]

When an invention developed at the NIH is licensed to a company, the NIH can receive royalties from the company. Royalties are based on sales, use, and value of the NIH technology to the development of the product. When it reaches the market, the NIH typically receives a small percentage of the profits.

In fiscal 2006, the National Institutes of Health and the Food and Drug Administration received $82.7 million in royalties, derived from 850 licenses.[67] This income represents about 0.3 percent of the annual NIH budget. The NIH can pay the inventors a share of the royalty income. Federal law limits to $150,000 the amount that individual investigators[68]—about 30 currently—can receive per year, and these payments can continue after the researcher leaves the NIH. Large amounts are exceptional, however. The average royalty that NIH investigators with patents receive per year is about $10,000.[62]

Salaries[7,69,70]

All employees at the NIH are paid according to salary systems established by the government. In general, compensation at the NIH for physicians, and particularly for investigators with only a Ph.D., is "highly competitive" with that paid at universities at the entry and mid-career levels, according to Colleen Barros, the chief financial officer. Nevertheless, it is at these levels that offers from medical schools and industry may succeed in luring people from the NIH. "They may not be doing as well as they wish or may want to be on a faster salary scale," explains Barros.[69]

Most of the 18,000 employees are paid according to the provisions of Title 5, in which salaries are assigned in 15 grades and 10 steps and range from $20,607 (grade 1, step 1) to $149,000 (grade 15, step 10).[71]

About 2,500 employees are paid according to the provisions of Title 42, which applies to the Department of Health and Human Services and, in particular, the NIH and the Centers for Disease Control and Prevention. Since Title 5 has no provision to pay foreign workers, many of whom are trainees and junior-level scientists, these employees are paid under the provisions of Title 42. Title 42 also allows employees to be paid more than does Title 5, and this includes institute directors, senior scientists, and clinicians who provide medical services in the Clinical Center. The top annual salary under Title 42 is about $275,000 plus the opportunity of receiving a bonus of up to $20,000, although few investigators make this much. Most scientific directors are paid between $200,000 and $250,000.

Clinicians in specialties where recruiting is particularly difficult because of competition from academic medical centers and private practice—anesthesiologists, for example—can earn as much as $350,000, as provided in Title 38, and may not be expected to do research. NIH recruiters talk of the easier professional life that doctors in the Clinical Center can live—less time on call, fewer operations per day, and no relationship of salary with how many patients they see. The NIH may sweeten a recruitment by paying a one-time incentive in return for a two-year service agreement and occasionally try to keep a leading investigator from being recruited elsewhere with a "retention incentive."[72]

Most investigators who leave the NIH for university life and continue to do research will have to write grants throughout their careers even after they achieve tenure, an obligation that NIH investigators, whether tenured or not, do not have.* This provision, in particular, is worth tens of thousands of dollars in salary to many NIH investigators.[73]

NIH benefits, however, are not as liberal as at many universities, where members of the faculty often receive financial help with college tuition for their dependents, plus other perks.[69] Not working for the government, furthermore, allows scientists to accept honoraria for giving talks, consulting, and writing books.

The NIH director, a political appointee, receives $191,700,† about the same amount as the Vice President of the United States.[69] The NIH director sets the salaries for each of the institute and center directors, "which vary, among other factors, by the size and complexity of their institutes and centers but are generally higher than that of the NIH director given their civil service nonpolitical status," according to NIH director Elias Zerhouni (2002–2008.)[74]

---

* NIH administrators refer to the grant-writing and "publish or perish" conventions at universities as "eat what you kill."[69]
† In fiscal year 2008.

CHAPTER SIX

Congress and Advocates

Congressional Friends

One of the NIH's greatest supporters in the Congress was John Edward Porter, Republican representative from the 10th District in northeastern Illinois from 1980 to 2001. Porter's father developed poliomyelitis as a infant and always wore a brace on his leg. "As a child, I thought that something was wrong with all the other fathers because they didn't have braces," Porter remembers.[1] He was very impressed when polio vaccines were announced in the 1950s. Exposed to science as an engineering student at Massachusetts Institute of Technology, which he attended for a year, Porter changed his career plans to the law—his father was a lawyer—and received his undergraduate degree from Northwestern University and his law degree from the University of Michigan Law School.

As a member of the Appropriations Subcommittee on Labor, Health and Human Services and Education for 20 years, Porter and his colleagues were responsible for assigning the funds for the NIH. When the Republicans became the majority party in the House in 1995, Porter became chairman of the subcommittee.

The Republican budget directed him to reduce all appropriations by five percent per year. "In all respects, but particularly in regard to the NIH and the CDC, I thought this was insane," said Porter. "Working with FASEB, [Federation of American Societies for Experimental Biology] we put together a group of Nobel Prize winners and businessmen to lobby Newt Gingrich [the Speaker]." After listening to the group, "Newt said that Republicans had made a mistake,"[1] and agreed to let Porter put the budgets for the NIH and Centers for Disease Control and Prevention (CDC) into a separate bill with an increase for NIH of 5.7 percent and a small increase for CDC. The Senate agreed, the legislation was passed, and Clinton signed it. "We'd protected the NIH from the budgetary wars," said Porter. "America should sustain its consistent investments in medical research and continue our worldwide leadership."[2]

The next year, Porter and his House colleague, Democrat David Obey from Wisconsin, and in the Senate Arlen Specter,[3]* Republican[†] from Pennsylvania, and Tom Harkin, Democrat from Iowa, got the NIH increases of "6.9 percent, 7.1 percent the following year, and then 15 percent for the next three years to begin the irreversible process of doubling the NIH budget, which was completed two years later," Porter explained.[2] An influential advocate for this result was former representative Paul Rogers (D-Fla., 1955–1979), then chairman of Research!America.[5‡]

The NIH budget increased by 15 percent a year, thereby doubling from about $13 billion to $26 billion from 1999 to 2003, the last two years of the Clinton administration and the first three years of the Bush administration.[7]

Advocates

Advocates for particular diseases can be a blessing and a curse for the NIH. While they convince members of Congress that they should allocate money for medical research, much of which will go to the NIH, their lobbying has funded the investigation of some diseases that the NIH would prefer being done elsewhere. Advocates are also held

---

* Specter was responsible for obtaining a large increase in the NIH annual budget when he provided a crucial Republican vote in the Senate to assure passage of the economic stimulus package in February, 2009.[4]

† Later Democrat.

‡ The recently constructed Building 35, the Neuroscience Research Center, is named for Porter, and the plaza in front of Building 1, the Shannon building where Zerhouni has his office, is named for Rogers. Paul Rogers died on October 12, 2008.[6]

responsible for the proliferation of the institutes and centers, which many believe increases administrative costs, encourages bureaucracy that limits the potential for undertaking risky, creative research,[8] and isolates scientists.[9] "Everybody wants an institute for his disease," Paul Rogers has observed. "The disease groups do the lobbying, and interested members of Congress respond to it."[10]

The attitude of the NIH directors to advocacy groups varies. Harold Varmus, when he began his tenure as NIH director in 1993, did not welcome their efforts,[11,12] but his attitude mellowed as he came to recognize their value.[12] Gerald Fischbach, former director of the National Institute of Neurological Disorders and Stroke, had heard that advocates would be a drag. "Not so," he said. "Though a few I talked with asked, 'If Congress is giving so much money to NIH, why am I still getting sick,' most understand the limits of medical science."[13] Advocacy groups have found that Elias Zerhouni has supported their aims from the beginning of his directorship.[12]

Many advocates, including members of Congress, champion particular diseases, often those that have afflicted themselves or members of their families. Understandably, they want cures, and quickly.[3,14] "What have you done for me?" is what many in the medical academic community think advocates are asking.[15] But that's not the way medical science works. "It's a faulty premise," said Barbara Culliton, a former writer for *Science* magazine. "They believe that the reason diseases are not cured is that we're not spending enough money. What is needed, that many in the advocacy community do not realize, are a few very good ideas that work out."[8]

Since 1971 when President Richard Nixon declared war on cancer, the disease has remained one of the two leading causes of death in the United States, despite all that has been learned, and even though the lives of many patients with cancer have been improved and extended.

In wanting cures right away, most advocates favor support for clinical over basic research. As an example, the National Alliance on Mental Illness (NAMI) supports funding for patients with schizophrenia and bipolar illness, rather than for less-ill people with mental disorders, according to Steven Hyman, former director of the National Institute of Mental Health. "They virulently oppose spending money for basic science and can't believe that we don't have the wherewithal to cure schizophrenia now."[16]

Mary Woolley, president of Research!America, disagrees. Her organization, according to its website, advocates "strong funding and policies that support the advancement of medical, health, and scientific research."[17] Woolley has found that many advocacy groups appreciate basic research and what it can do. She is convinced that the growth in NIH funding was due to "the patient-advocate community...insisting and not because scientists pushed for it. The science community just want[s] advocates to be quiet and send money."[18]

Another group with a similar mission is the National Health Council, to which many health-related organizations belong. "We represent all patients with chronic diseases with a single voice and fight for increased funding for the NIH," explains Myrl Weinberg, the council's president. "We don't lobby for particular diseases."[12]

Advocacy for biomedical research has become more professional. Whereas, until the 1990s, most health-supporting advocates were concerned citizens often associated with nonprofit foundations, now the organizations employ professional lobbyists to champion their diseases. Critics say that NIH programs can become skewed by lobbyists, who are paid to deliver for the organizations that employ them, not to think broadly about policy.

Advocates significantly helped bring about the doubling by persuading Congress to support the NIH. When flat budgets arrived, the effectiveness of their efforts to increase NIH funding within those budgets decreased. Many representatives and senators seemed to have grown tired of listening to medical advocates and their lobbyists, sometimes accompanied by "poster patients" who pestered them in their offices. "Additional money wasn't there any more, and advocacy fell apart," Mary Woolley said. "Congress became distracted by other issues: conflict-of-interest, misuse of appropriated funds, huge other agendas [including bioterrorism³] and, in effect, said to the NIH, 'Get over it.'"[18] Don Gibbons, a public affairs official from Harvard, felt that "people on the Hill were having NIH fatigue."[19] NIH advocates believed that many in Congress were telling the research community that they'd done their job: "now go away."[18-21]

Group of Concerned Universities and Research Institutions

In 2006, in response to the financial pain being felt by all research-intensive academic medical centers, Harvard University organized a consortium of leading institutions to lobby Congress in support of increased NIH funding[19,22-24]* since NIH, being a government agency, cannot lobby for itself.[1] NIH employees, however, can provide data in response to outside requests.

The Group of Concerned Universities and Research Institutions will provide Congress with a more effective message about the effects that flat funding was having on scientific progress than have other, similarly motivated, groups. Those involved in the creation of the consortium had come to believe that a special joint venture was necessary to supplement umbrella groups like the Association of American Medical Colleges (AAMC), Federation of American Societies for Experimental Biology, and the American Association of Universities. These organizations were often encumbered by other responsibilities, competition among the medical schools that make up the societies, and lack of coordinated leadership from the officers with their one-year terms of office.[22] The members of the Group hoped to overcome the tendency of university officers, when asking for money from Congress, to lobby specifically for their institutions and not for the NIH.

Some, however, were willing to appear before Congress. On March 19, 2007, Joan Siefert Brugge, chair of the department of cell biology at Harvard Medical School, testified before the Senate Labor, Health and Human Services subcommittee of the Appropriations committee. She spoke about the difficulties investigators like her were having obtaining funding from the National Institutes of Health; in her case, the National Cancer Institute. In emphasizing how much more competitive the chances of getting grants had become, she said to the members of the subcommittee, "Getting rejected for funding for a grant in the 10–20 percentile range [a very high competitive score] is equivalent to flunking an exam with a grade of A."[25]

"It's a very difficult time," said Joseph Martin, dean of the Harvard Medical School from 1997 to 2007. "Faculty feel negative with funding so hard to get. Why should doctors choose research when the age of investigators getting their first R01 grant is now 42 to 43 years,"[23] a number that the NIH confirms.[26] The average age when medical scientists at U.S. medical schools receive their first tenure-track

---

* The universities and hospital groups in this consortium are California, Columbia, Harvard, Johns Hopkins, Partners HealthCare [Massachusetts General and Brigham and Women's hospitals], Washington University in St. Louis, University of Wisconsin-Madison, Texas, and Yale.

appointment, Zerhouni has written, is 38, and their first independent NIH grant, age 42.²⁷

Although deeply concerned about how the NIH's current budget is affecting Harvard, Martin worried that the financial problems of the NIH will selectively affect schools that are not now in the research-intensive category. "Some of their investigators will have more and more trouble getting funded, and unless other sources, such as the state government in schools owned by states, can make up the difference, this maldistribution will be amplified."²³ Schools, both private and state, that were able to increase their research productivity when the NIH budget doubled in five years, may find themselves less able to concentrate on research.

"We're likely to see an even more drastic dissociation between the 'haves,' the top 20 to 25 schools, and the 'have-nots,'" said Martin.²³* Another dean from a research-intensive school fears that the research-deprived schools could become "more like trade schools." When the policy of doubling the budget was developed, "the Clintons were in the While House, there was a budgetary surplus, and intellectual imagination. Now there are wars and a moral crisis."²⁸

Martin is not optimistic about the near future. "There's no room to wiggle," he believes. "The wars in Iraq and Afghanistan, the budget deficit, the growing impetus for the federal government to contribute more financing to health care, will take priority."²³ John Marburger, President Bush's science adviser, stated as much in an interview with *Science* magazine in May, 2007. "I haven't seen any evidence of an increased top line for science ... I think that's wishful thinking."²⁹

A Democratic administration and Congress will face the same limitations, Martin predicts, no matter how much its leaders may want to better support the NIH. Although some states have taken on the responsibility of financing specific research—studies with stem cells in California, for example—he fears that this may lead to "de-federalizing of funding for certain research areas."²³ "It's not fair, since not all states give these grants," said Harold Varmus. "One of our strengths is having a level playing field. It's worrisome."³⁰

---

* For more about the implications of research-rich versus research-poor universities and colleges, see: Richard Lewontin, "The Socialization of Research and the Transformation of the Academy," in C. Hannaway, ed., *Biomedicine in the twentieth century: Practices, policies, and politics*. Washington, D.C.: IOS Press, 2008, pp. 22–23.

One observer predicted a grim future. "Even an eight percent increase in the budget will not bring things back to the glory days, because there are so many more people applying and expenses are so much higher. The money lost during the years of lower budgets is gone. It cannot be made up."

CHAPTER SEVEN

Directors

AFTER THE REIGN OF James Shannon, the iconic[1] director who ran the NIH with great authority and success from 1955 to 1968—"the golden years" as NIH historian Victoria Harden calls them[2]*—directors have been able to exercise considerably less power than the directors of the institutes. The institute directors receive their budgets directly from Congress and, unlike the director, until recently, award grants. The NIH director has a small budget for his office and has to be responsive to the Secretary of Health and Human Services and the White House.

After Shannon, the NIH director and the institute and center directors formed a "council of equals," though not quite so egalitarian in the case of some NIH directors, as Victoria Harden describes the relationship. "Just look at the floor covering in the halls—vinyl in Building 1 [where the director's offices are], carpets in the institutes."[2]

"The NIH director's not that powerful," said Marc Kirschner, a professor at Harvard. "His chief job is appointing institute directors."[4]

Observers both inside and outside the NIH consider the NIH directorship a difficult and lonely job,[5-7] greatly influenced by political considerations.[8] The political affiliations of two of the last three NIH

---

*See Buhm Soon Park's enthusiastic description of how Shannon built the early scientific programs at the NIH in *Perspectives in Biology and Medicine* 2003;46:389–394.[3]

directors were open to the public and Congress. Bernadine Healy (director from 1991–1993, George H. W. Bush, President) was an enthusiastic Republican who had worked in the Reagan White House and would run for the United States Senate after leaving the NIH. Harold Varmus (director 1993–1999, Bill Clinton, President) had revealed his political orientation by joining a group of scientists supporting the Clinton-Gore ticket in 1992.[9]* The most recent director, Elias Zerhouni (appointed in 2002, George W. Bush, President) has hidden his politics from view, a practice that Kenneth Shine, former president of the Institute of Medicine, applauds. "I think directors need to be very careful to walk a non-partisan course."[11]†

Many institute directors consider their jobs the best ones at the NIH for people skilled in academic administration, and some have declined to become NIH directors when offered the post. Anthony Fauci, director of the National Institute of Allergy and Infectious Diseases, is the example all point to. Highly respected for both his scientific and his administrative skills, "Fauci should have been NIH director," believes Steven Hyman, former director of the National Institute of Mental Health. "He's a thoughtful anchor of NIH and politically savvy. You want him on your side."[13] However, Fauci rejected the job twice when offered the position during the administrations of George H. W. Bush and Bill Clinton.‡ "You can't do science from the director's office," he explained. "I didn't want to get involved in things I wasn't interested in."[6]

Comparing the doubling with the flat budgets that Elias Zerhouni faced during part of the time he was director, Fauci said, "The circumstances you find determine your tenure."[6] What doesn't change, however, "is the core mission," said a former institute director. "Every leader wants to establish his own imprint."[14]

The recommendations of the Institute of Medicine's (IOM) 2003 report[15,16] on the NIH could improve certain deficiencies of the office of the director that have discouraged some candidates from accepting it. Since the director has few funds of his own,[8,17] the report advised that his office be given a larger budget "to conduct NIH-wide planning

---

* Subsequently, Varmus led a science advisory committee established by Barack Obama when he was a candidate for the presidency.[10]

† Toward the end of the second Bush administration, Zerhouni stated his opposition to the administration's policy on stem-cell research.[12] See Chapter 7.

‡ The George W. Bush administration also considered Fauci for the job. (See below.)

for trans-NIH initiatives[18].... The [IOM] committee believes that too little weight has been placed on potentially distinctive contributions of the IRP [intramural research program].[15]" The report also suggested that the directors serve a six-year term unless removed by the President. After a positive external review of their performance, directors could serve a second and final six years on recommendation of the Secretary of the Department of Health and Human Services. Despite these changes, "institute directors have the best job," said former genome institute director Francis Collins.[5]

Office of the Director

The Office of the Director (OD), according to the NIH website, "is the central office at NIH. The OD is responsible for setting policy for NIH and for planning, managing, and coordinating the programs and activities of all the NIH components."[19] Since 1995, the NIH director has reported directly to the Secretary of the Department of Health and Human Services and no longer to the Assistant Secretary of Health.[20] In 2008, the budget for the OD was $1.112 billion, of which $498 million was assigned to the Common Fund, which supports the NIH director's initiatives and includes the Roadmap for Medical Research.*

Bernadine Healy

Twenty-one months elapsed between the departure of James Wyngaarden, director of the NIH from 1982 to 1989, and the arrival of Bernadine Healy as the eighth director† of the NIH on April 9, 1991. The interregnum coincided with the end of the Reagan and the beginning of the George H. W. Bush administration, which was challenged to find someone politically and scientifically acceptable who would take the job.

Anthony Fauci had declined Bush's offer to become the NIH director. "I wanted to continue my direct leadership of the AIDS research effort both as an active scientist and as the director of NIAID. I told this directly to the President in the Oval Office in the

---

* See Chapter 7.
† From when the name of the National Institute of Health became plural in 1948 as the National Institutes of Health.

presence of John Sununu, and the President graciously accepted my decision."[21]

Healy, the doctor who was eventually selected, had more political experience than most academic cardiologists have had (or may wish to have). She graduated from Vassar College and Harvard Medical School and trained in medicine and cardiology at the Johns Hopkins Hospital and at the National Heart, Lung, and Blood Institute, where she was a clinical associate. Healy then returned to Hopkins, joined the medical faculty, became director of the coronary care unit at the Johns Hopkins Hospital, and rose to professor of medicine and assistant dean at the medical school.

Healy began her non-academic political career in 1984 as deputy director of the Office of Technology Policy in the Reagan White House. She left the government a year later to become director of the Lerner Research Institute of the Cleveland Clinic.[22]* In 1988 and 1989, she was president of the American Heart Association, the specialty's leading professional and lay organization.

In 1990, as she tells it, "I first heard about the NIH job from a newspaper story reporting that the search committee had included me as one of five candidates to succeed Jim Wyngaarden as director."[24] The leading candidate, Anthony Fauci, had refused the position. Louis Sullivan, Secretary of the Department of Health and Human Services, then offered Healy the job, and she accepted.

One significant hurdle Healy had to surmount involved the first Bush administration's policy of banning government support for research on fetal tissue on the assumption that such research might foster abortions.[25]† Healy had served on an NIH panel studying the issue and did not subscribe to the administration's position.

While a member of Reagan's government, she had been told, "Remember, you're a guest in someone else's living room." So, Healy explains, "I was prepared to suppress my personal opinion on the fetal tissue matter as I had done on the abortion issue under Reagan. If you're in a presidential appointed position, you can't oppose his political agenda."[24]

Healy was impressed that she was chosen for the NIH job despite her stated opinions on controversial issues. "I never backed away from my

---

* Where her husband, Floyd Loop, was chairman of the department of cardiothoracic surgery. Loop would become CEO of the Cleveland Clinic Foundation in 1989.[23]
† In 1993, President Bill Clinton would repeal the ban on federal funding of fetal tissue transplantation research.[26]

pro-choice and fetal tissue positions." Healy, despite her political affiliation, describes herself as a "screaming liberal" on some issues. A reporter in *Science* magazine wrote that she said, "I'm a feminist. . . . That's the amazing thing."[27]

Healy found the senior President Bush very supportive of the NIH, and her successor, Harold Varmus, praised Bush's support of science in general.[28] The president and his wife attended Healy's swearing-in.[24] "The administration clearly honored my private thoughts," she said.[24]

Healy's Projects

Women's Health Initiative[11,29–34]

Healy recognized from her work in academic cardiology that large clinical trials in her specialty seldom included many women in the sample. Accordingly, she launched the Women's Health Initiative during her first year as director. Barbara Mikulski, Democratic senator from Maryland, and Pat Schroeder, Democratic representative from the first district of Colorado (1973–1997), had encouraged the NIH to establish an Office of Research on Women's Health[35] in the early 1990s.

The most widely known program was a 15-year project—the longest the NIH has ever sponsored—that involved more than 161,000 women, aged 50 to 79. The study aimed to provide information about the effects of hormone therapy, dietary patterns, and calcium/vitamin D supplements on women's health, and specifically, on whether these variables would prevent heart disease, cancer, and osteoporosis.

Healy, who called the initiative "a moon walk for women," knew it would be expensive, eventually costing $650 million over the period of the study, and controversial. "It was controlled from the director's office," she said; "centrally planned—like the Human Genome project that I also strongly supported—not the usual NIH way [through the institutes]."[24]

Though critics predicted that hormone therapy would protect women's hearts—"a slam dunk" and not worthwhile to study, Healy was told—the data unexpectedly[11] showed that the drugs did not protect women older than 60 years of age from developing cardiovascular disease.[24]

James A. Shannon Director's Awards

These are one-time grants for investigators whose scores just miss the level for funding, but whose work a special inter-institute committee sees as particularly creative and worthy of support. Healy established them—"my MacArthur Awards," she called them[24]—in honor of James Shannon, the legendary NIH director (1955–1968), whom NIH historians and others with long memories credit with creating the NIH in its contemporary form.[36]* These grants do not bring overhead with them, to the annoyance of university officials who rely on these supplementary payments to support the infrastructure of the institutions where the awardees work.

Strategic Plan

Healy wrote a strategic plan for the NIH that she saw as providing a comprehensive "investment in humanity. The NIH would do good science but good science on behalf of humans."[24] Among other features, her plan would centralize the planning of some of the extramural research rather than relying as much on investigator-initiated projects.[37] She also proposed that the NIH be given departmental status rather than continue under the Department of Health and Human Services. This plan, promoted from time to time by other senior NIH officials, failed again to be implemented.

Healy wanted to raise public recognition of the NIH,[38] though in her short term as director, little was accomplished to change this. She was concerned that the NIH "did not seem to be known 'outside the Beltway' as well as agencies like the CDC [Centers for Disease Control and Prevention] or NASA," remembered Griffin Rodgers, director of the National Institute of Diabetes and Digestive and Kidney Diseases.[39]

Healy favored the NIH's patenting human genes discovered by its staff. "The government should obtain a return on its investment by patenting what we discover," she explained.[24] The plan was strongly opposed by many scientists[40] who believed that DNA belonged to the public.[37] Nobel winner James Watson, the first director of the federal Human Genome Project, called patenting genes "lunacy."[41] J. Craig Venter writes that he was then leading the gene-sequencing project at the NIH, made the gene discoveries that were the subject of the NIH patents, and had no choice over patents being filed even though he was against the patenting.[42]†

---

* Building 1, which houses the offices of the director and other NIH administrators, is named for Shannon.
† For more details, see Venter's autobiography, *A Life Decoded. My Genome: My Life* (New York: Viking Press, 2008).[43]

The strategic plan, about which Healy felt very strongly—as she did about most of her endeavors—was never instituted.[44] Colleagues observed:[2,7,33]

- "People thought the NIH didn't need a strategic plan. 'Just fund my research,' they would say."
- "It fell between those who like to plan and those who were solely committed to investigator-initiated research."
- "It was published when she left but was dead on arrival. It sits on a shelf somewhere," said Barbara Culliton, then the news editor at *Science* magazine. "There was more anxiety about what she was up to than what she actually did."[37]

Opinions[2,6,7,11,14,25,26,37,45–64]

Primarily a clinician rather than a bench scientist, and the first woman director of the NIH,[49] Healy was "not in the old boy's club," remembered NIH historian Victoria Harden.[2] Her family did not move to Bethesda with her, so she commuted on weekends to Cleveland, where her husband and children lived. During the week, she "rattled around"[2] in the big house* on the campus that the NIH provides for the director.†

Like Donald Fredrickson (director from 1975–1981) and James Wyngaarden (1982–1989) before her and Harold Varmus afterwards, Healy chafed at her lack of authority when compared with the power exercised by the institute directors.[33,49] Like some of her predecessors and successors, she tried to change this but, also like them, she couldn't.

Scientifically, Healy was recognized as "very bright" and praised for making some excellent appointments. She recruited Kenneth Holden as the third director of the National Institute of Environmental Health Sciences,[20] chose Michael Gottesman as acting director of the Genome Program, and led the recruitment of Francis Collins as the permanent director—"a master stroke," as one of the NIH veterans said.

---

* The NIH campus at Bethesda contains 14 colonial-style homes built between 1939 and 1941 when NIH moved there.[47]
† NIH historian Victoria Harden sees symbolism in Healy's portrait in the collection of pictures of all the directors that march down the hall on the first floor of Building 1. All but one of the men wear suits, shirts, and ties. Healy appears in a fashionable suit with a picture of her family and the American flag.[2] Harold Varmus is tieless and coatless with his shirt collar open. See below.

Persuading her medical school classmate Richard Hodes[65] to leave the National Cancer Institute to become director of the National Institute on Aging was also applauded.

Though commended for several of her projects, especially the Women's Health Initiative and the Shannon awards, Healy was criticized for being, at times, imperious, unaccepting of new ideas or criticism, and "so ham-handed she defeated" some of her best ideas. People contended that she was trying to change things too quickly at an institution whose "core doesn't torque easily." Several critics said that she didn't obtain the "full confidence of the scientific community." Although "she presented the public side well," some of her relations with Congress were troubled.

Administratively, she relied strongly on her staff rather than on the institute and center directors, which some criticized as inconsistent with the customary patterns of the federal bureaucracy. Her style of leadership was "top-down."

"A blunt-talking New Yorker born and bred in working-class Queens, she was not known as a diplomat," wrote a reporter in a profile of Healy in *The New York Times*. "Rather, she was known as a driven professional who ruffled feathers but made things happen."[66]

Healy was also criticized for her strong support of the Republican Party in an institution where, it is said, liberals and Democrats predominate. "She saw the job more as a political appointment than the other directors had and tried to make the place more hierarchical," said one observer. "She raised her political head too high," said another, and this helped to account for her losing the job when the administrations changed.

Hoping to continue as director,[2] Healy did not submit the traditional resignation letter to the White House when the administrations changed.[26] George H. W. Bush's successor, Bill Clinton, however, wanted someone else to direct the NIH, and accordingly, Donna Shalala, the new Secretary of the Department of Health and Human Services, refused to continue Healy's appointment.[24*] It was rumored that Healy had discussed the possibility of becoming the vice-presidential candidate in Ross Perot's unsuccessful presidential bid in 1992. If true, one observer said, "this would not have endeared her to either Bush or Clinton."

---

*For further details about Healy's departure, see *Science*, vol. 259, 5 March 1993, pp. 1388–9.[47]

Post-NIH activities

After leaving the NIH, Healy unsuccessfully sought the Republican nomination for the United States Senate from Ohio in the 1994 general election. From 1995 to 1999, she was dean of the College of Medicine and Public Health at Ohio State University and then president of the American Red Cross until 2001. She is now a columnist and health editor at *U.S. News and World Report*.

Harold Varmus

Harold Eliot Varmus* became the fourteenth director of the NIH on November 23, 1993.† He came from the University of California–San Francisco (UCSF), where, with colleague J. Michael Bishop, he had won the Nobel Prize in Physiology or Medicine in 1989. He was nominated for the NIH job by a committee chaired by Bruce Alberts, president of the National Academy of Sciences and a colleague of Varmus at UCSF.‡ Donna Shalala said that she wanted "a world-class scientist who would exemplify scientific excellence"[9] as the next NIH director. She got one.

Varmus heard that some were troubled that he had never been a department chairman and didn't have the administrative experience for the job. "The biggest thing I'd run was a laboratory," he

---

*Grew up in Long Island, New York, son of a general practitioner (father) and a psychiatric social worker. B.A., Amherst; M.A., Harvard (in English); M.D., Columbia; intern and resident, Columbia-Presbyterian Medical Center; clinical associate NIH. See Varmus's superb memoir, *The Art and Politics of Science* (Norton, 2009),[67,68] in which he describes his experiences as NIH director, some of which I do not cover in this book.

†During the interregnum following Healy's departure, Ruth Kirschstein, most recently director of the National Institute of General Medical Sciences, became acting director. Varmus would name her deputy director during his term as director. "Ruth had set the model for institute directors," said Alan Leshner, then director of the National Institute on Drug Abuse. "Retaining her as deputy director with Harold was terrific. She had faced all the bureaucratic issues he would face. Some healing was needed after Healy."[69]

‡The two other people on the short list were Judith Rodin, then provost of Yale University and soon to be president of the University of Pennsylvania, and William Danforth, M.D., then chancellor of Washington University in St. Louis, "who agreed to have his name submitted in case the White House had a problem with Varmus or Rodin," according to Philip Lee, Assistant Secretary for Health in the Department of Health and Human Services at the time.[26]

acknowledged. "I never wanted to be a chairman, didn't think I'd like the administration." At the time the NIH was employing more than 17,000 people and had a budget of $11 billion.[70] "So I checked about the administrative backup I would have."[28] Varmus concluded that it was excellent, and, when offered the job, he accepted.[28]

Others brought to the NIH from universities agree with Varmus. Gerald Fischbach, whom Varmus was recruiting to become director of the National Institute of Neurological Disorders and Stroke in 1998 found the administrators at the NIH superior to those he knew at Harvard. "I found two types, each excellent—the science officers, many of whom came from academics or business, who think about science but don't do it, and the others, who make the trains run on time."[71] Guy McKhann, another university-based academic neurologist who had briefly worked in a senior position at one of the institutes, and William Kelley of the University of Pennsylvania and a clinical associate there, agree.[7,72] "The NIH is extremely well run for a government-run organization," Kelley said.[7]

Unlike his recent predecessors, Varmus set up a small research laboratory at the NIH, initially staffed by five of the 25 people who had worked with him in San Francisco. He was the first NIH director to run a research laboratory since William Henry Sebrell was the director (1950–1955). However, Varmus explains, "to say that I continued to work in the lab will mislead many. I supervised trainees and technicians but was never at the bench."[20]

Varmus encouraged institute directors, few of whom at the time still ran research programs, to continue to work in their laboratories.[65] He believed that institute directors, scientific directors, and clinical directors do their best administratively when they continue doing research,[71] a point of view shared by many of the institute directors who retain laboratories,[14,65] though not by all of them. As for his own research while director, Varmus said that his scientific productivity could have been greater. "Actually, I had a very good lab experience.... Without [it] I think I would feel pretty starved for scientific conversation. I would like to have had a little more time to read scientific literature."[70]

One longtime NIHer cites potential conflict of interest when high-level scientific administrators run their own labs.[31] Scientific directors whose large personal research programs reduce their commitment to the administrative chores that come with these appointments can lose their jobs. In recent years, three scientific directors and one institute director were asked to leave, partially because of this problem. Each had been recruited from outside the NIH and, consequently, they were not conversant with the customs and regulations that veteran intramural investigators observe.[73] Alan Schechter, a physician-scientist with 40 years of experience at the NIH, said, "There is now a consensus among those with whom I talk that recruiting scientific directors from outside is very risky."[*]

A Transplanted Westerner

Varmus and his Californian informality stood out in the Washington culture of conservative suits.[49] Although he could produce a jacket and even a tie,[†] which he kept in his office when presidents and other notables visited the NIH, on most days he wore frayed khakis and open-necked shirts.[9][‡] Eschewing the city's car culture, he rode his bicycle the 12 miles to work each day from his home in Washington, where he and his wife lived in preference to the director's house on the NIH campus.[47] After taking up sculling, he'd bike four miles to the boathouse, row for 45 minutes, then ride on to the NIH. The route back home took him through Rock Creek Park. "Washington was great for my physical training," he told a reporter.[70][**]

In his profile of Varmus in *The New Yorker*, James Fallows wrote:[9][***]

In person, Varmus is a lean, energetic, and intense presence.... He is just over six feet tall, with strongly muscled legs from decades of serious bicycle riding, and he can be seen as loosely hinged as a teenager—slouching in a chair during a meeting, jiggling coins in his pocket when he stands. He speaks rapidly, with the precision of a scientist, but also with a bemused smile and a string of dry

---

[*] "One great success in the last decade," Schechter adds, "is Betsy Nabel, who came as scientific director for clinical research in NHLBI [National Heart, Lung, and Blood Institute]" and is now the director of the institute.[74]

[†] Which were described as "ill-matched."[57]

[‡] Accordingly, Varmus was not wearing a tie when I interviewed him in a New York restaurant on a Saturday in March, 2007—but then neither was I—or in his office at the Memorial Sloan-Kettering Cancer Center in January, 2008.

[**] The first time he bicycled to work, NIH guards told Varmus that he couldn't bring his bike into Building 1 where the office of the director is located. This restriction did not apply for long.[57]

[***] June 7, 1999, issue.

wisecracks.... like Ronald Reagan, whom he resembles in no other way, Varmus seems totally relaxed in his public role, enjoying the enormously complicated process he oversees daily.

"It became all about science"

The scientists felt they had achieved nirvana with a Nobel Prize winner as the director.[5,6,14,37,56,58,75] With the doubling of the budget and the presence of a Nobelist as director, many considered this period a second "golden age at the NIH."[69]

Varmus was committed to recruiting only the most talented scientists and would leave positions open until he could land the right people. Kenneth Shine, then president of the Institute of Medicine, thought this to be one of his "greatest contributions."[11]

"The emphasis was on quality," says Alan Leshner, whom Varmus had selected as director of the National Institute on Drug Abuse. "He was less interested that the candidates be strong administrators."[69]

"Sharpest mind I've ever met," says Francis Collins, who led the NIH component of the Human Genome Project. "Just talk to him for a few minutes, and he'll know more than you about what you thought you knew."[5]

Leshner agrees. "Meeting with him was like your doctoral review. He knew about everything, and wouldn't waste any time."[69]

"A brilliant, terrific guy," is the opinion of Anthony Fauci. "As you'd expect, he learned very quickly. It was a trauma-free tenure with no scandals, and he was very popular with Shalala and Clinton."[6] In the profile on Varmus in *The New Yorker,* Shalala said, "In general in public life, appointing one individual doesn't make a huge difference. But this appointment was absolutely crucial. It may turn out to be the most important legacy of the Clinton Administration."[9]

"It became all about science," Patricia Grady, the director of the National Institute of Nursing Research, says. "Harold re-energized our mission."[76] Richard Hodes, director of the National Institute on Aging (NIA) declares; "he made science the primary language and currency at the NIH."[65]

Several senior leaders say that Varmus exhibited more administrative ability than expected, given his limited previous organizational experience,[6,14,52,69,76] "and he has not been faulted by the scientific community for the NIH's current situation."[77] In spite of his having spent much of his career in the laboratory, Varmus was seen as a "political animal."[33]

- "He had to learn leadership, which is an acquired skill. He applied his brain to this with the same seriousness he applied to his science."[69]
- "By cutting some of the red tape, he reduced the number of bureaucratic steps necessary to get things done, and made hiring easier."[76]
- "He's a consummate director and excellent manager but not a micromanager."[55]
- "Despite his brilliance, he listens to people."[57]

Varmus was very effective in extolling the virtues of the NIH to the public and to members of Congress.[11,26] "Congress loved him," says one of the center directors.

Varmus became director when the NIH budget was flat or sinking.[78] Corrected for inflation, the budget had dropped from $16.3 billion to $16.2 billion from 1994 to 1995.[79] Then came the doubling—which Varmus had advocated before he became director[58,80]—and Varmus could concentrate on recruiting institute directors of outstanding quality,[55,72] although a few recruited from the outside would leave after relatively short terms.[4,55]

To penetrate the intellectual walls surrounding each of the institutes,[48] a committee recommended the creation of "Scientific Interest Groups," assemblies of scientists with common research pursuits,[81,82] a project that Varmus and Gottesman encouraged.[20] The interest groups sponsor symposia, poster sessions, and lectures; offer mentoring and career guidance for junior scientists; help researchers share the latest techniques and information, and act as informal advisors to senior officers. Many of these groups are co-sponsored by neighboring academic and government institutions and welcome interested non-NIH scientists.

The scientific interest groups exemplified Varmus's desire to make the NIH more collegial. He encouraged collaboration between institutes. "You have to play nicely with everyone in the sandbox," he was heard to say.[69]

Praise and Criticism[2,5,6,14,29,31,46,48,49,52,54,55,57,58,65,69,72,76,81,83–88]

Not all at the NIH, however, appreciated everything Varmus accomplished. "I think that some of his policies were not a good thing for the NIH in the long run," said Alan Schechter, a veteran NIH investigator in the National Institute of Diabetes and Digestive and Kidney Diseases. Schechter, among others,[29] regretted what they saw as Varmus's "lack of interest or belief in clinical research. He appeared to believe that you learn more about cancer from studies in mice than in people."[25] Though Varmus supported funding clinical research in the extramural community, he initially favored molecular and genetic research in the intramural program.[29,86] "But he didn't ignore clinical research," said one of the institute directors. "He told me to make ours better."[69]

Most of the people at the NIH interviewed for this book were very supportive of Varmus's tenure as director.[72] "Harold underwent a transformation here," said Harvey Alter, the distinguished NIH investigator. "Surprisingly, he became a born-again clinician and concentrated on integrating laboratory to clinical research."[83] Varmus was widely praised for improving the morale of those working in the intramural research program.

Not unexpectedly, some found Varmus, like Healy, "arrogant" and "peremptory," though in his case "the driving force was intellect and accomplishment more than character," said Barbara Culliton.[37] His scientific accomplishments and personality made some of his colleagues see him as intimidating.

"Not everyone liked him. He doesn't like to schmooze, and some of his social skills are not the best," said Alan Leshner, who greatly admires Varmus. "He's a very high pressure guy, but he takes people as they are. When Varmus tells you about a problem in your unit, he'll finish with, 'I told you about the problem. Now let's not discuss it further.'"[69]

Varmus chafed at some of the constraints of working in Washington's politicized scientific world. "Harold never wanted people to write his speeches in the approved governmental way," said NIH historian Victoria Harden. "He just wasn't used to getting every word checked."[2] Though "mild on first contact, he turned out to be very forceful," according to Milton Corn at the National Library of Medicine. "He seemed to have little trouble dealing with the institute directors. Publicly, he *was* the NIH."[46]

"Varmus revitalized the place. People were proud to have him there," said Philip Lee, a former Assistant Secretary of the Department of Health and Human Services and longtime member of the faculty at UCSF. "He avoided getting pulled into the Clinton health care reforms."[26]

Varmus Leaves

In 1999, Varmus left the NIH to become president of the Memorial Sloan-Kettering Cancer Center in New York, to the regret of many of

Chapter Seven: Directors

his NIH colleagues. Francis Collins, for one, was very sorry that he left, but understood that, "Harold loves New York and working on cancer."[5]

"By the time I left the NIH, I was disillusioned," Varmus explained. "I wanted to run something but also spend more time in my lab than I could at the NIH. At Sloan-Kettering, I've been able to hire more scientists, bring Memorial Hospital closer to our research institute, and build science internationally."[28] In his new job Varmus has emphasized translational research and has encouraged the cancer genome project. He finds New York City "much better now [than when he was growing up and going to medical school there]. Being here gives one great entrèe. The fund-raising capacity is great"; though, he said, that activity takes only about five percent of his time.[28] Nevertheless, his fund-raising has proved quite successful, providing the money to build a 23-story research tower for Sloan-Kettering. Varmus divides his time almost equally between running Memorial Sloan-Kettering, working in his research lab, and engaging in outside scientific-political activities.*

Between the Varmus and Zerhouni Directorships

The Clinton administration tried to find a successor to Varmus during its last year but had to step back,[33] concerned about triggering a lengthy Senate confirmation hearing over candidates' views on whether the NIH should finance research on human embryonic stem cells.[90]

Gerald Fischbach, director of the National Institute of Neurological Disorders and Stroke, had been one of the Clinton administration's nominees. The FBI was vetting Fischbach, and he had begun visiting influential members of the Congress. "Then it suddenly ended," he remembers.[91] Fischbach had testified before senators Specter and Harkin, both friends of the NIH, about embryonic stem cells, and had said that not being able to work with them was like "doing research with one's hands tied behind one's back."[91] One of Fischbach's interests is Parkinson's disease, for the treatment of which stem cell research shows promise.† Senator Sam Brownback, Republican from Kansas,

---

*In December, 2008, Barack Obama appointed Varmus a co-chairman of the President's Council of Advisors on Science and Technology.[89]
† See Chapter 9 on the stem cell controversy.

and other opponents of federal funding for stem cell research also gave evidence.

"The Clinton administration had decided not to push my nomination and not on the merits of the case. It was a political decision," Fischbach remembers. "I wasn't pleased. I would have liked to do it. The NIH directorship is the most important job in medical science. The NIH does God's work. It's all about science and health."[91]

For the remainder of Bill Clinton's term, from December, 1999 to January, 2001, and for more than one year into the George W. Bush administration, the NIH had no permanent director. Ruth Kirschstein, then NIH deputy director, became interim director at a time when the budget of the agency was growing by 15 percent per year. Many were concerned that no plans had been established for this unique opportunity and noted that the budget was simply doubled for all institutes and disease areas.[92]

Bush and Tommy Thompson, his Secretary of the Department of Health and Human Services to whom NIH directors report, delayed starting a search for a new director until the President announced his stem cell policy in August, 2001.[92] Then the administration had to locate a suitable nominee who was sufficiently eminent and subscribed to the Bush medico-social agenda. Finding someone was not easy as so many members of the scientific community had supported Al Gore in the 2000 election.[11]

Anthony Fauci again became a leading candidate for the job. Fauci explains what happened:[21,93]

> Between the terms of Varmus and Zerhouni, both Tommy Thompson and President George W. Bush wanted me to become Director of NIH. At that point I made it clear to Tommy Thompson that I would do it only under the condition that I would be allowed to simultaneously be Director of NIAID and Director of NIH since I did not want to give up my direct leadership of the AIDS and Infectious Diseases Research effort. [Thompson] offered me the NIH job and said that it was fine with him if I also did the NIAID job. Obviously, he would have to pass this on to the White House for their approval and official offer.... [The White House] explicitly [said] that if you want the NIH job, it is yours, but you cannot have both the NIH job and the NIAID job. If you do not want to

Chapter Seven: Directors

give up NIAID, then we cannot officially offer you the NIH Director job. I chose the NIAID.*

According to *The Washington Post*, Thompson "said that the White House had rejected Fauci, a move for which conservative activists took credit because Fauci was deemed 'insufficiently pro-life.'"[94] Fauci doubts the significance of this:[21]

> Senator Brownback (R-Kansas) was particularly concerned and even questioned me about some articles that I had written years earlier about the potential of using fetal tissue-derived lymphocytes in an attempt to reconstitute the immune response of HIV-infected individuals. However, notwithstanding *The Washington Post* article, I really do not think that Brownback or any ideological issues had any real impact on the decision regarding the NIH Director. I believe strongly that the issue was related to concern on the part of some in the White House about my doing both the NIAID and NIH jobs at the same time.

"I talked with Bush," said Ruth Kirschstein, who, for the second time, was interim NIH director. "'I bet you can't wait for me to get someone,' he told me. I told him that I wanted to make sure that we got a good scientist as the director."[33]† In part because of the hiatus in appointing a director, five institutes had no permanent directors when the new appointee was finally confirmed.[95]

Elias Zerhouni

The candidate chosen was Elias Zerhouni, then at the Johns Hopkins University School of Medicine as executive vice-dean and director (chairman) of the department of radiology. Zerhouni became the fifteenth director of the NIH on May 2, 2002, two years and four months after Harold Varmus left for Sloan-Kettering.[95]

---

* Phillip Lee said that Fauci's keeping the NIAID directorate while serving as NIH director "would have been a real conflict of interest."[26]
† Ruth Kirschstein died on October 6, 2009. She had worked at the NIH for more than 50 years.[94a]

Zerhouni had been told that the administration would not press him about the stem cell controversy if he kept his opinions to himself.[2] Zerhouni, who had been instrumental in establishing the stem cell research program at Johns Hopkins when he was executive vice-dean there,[96] did not publicly oppose the administration's policy against government support of stem cell research at the time of his appointment. *The Washington Post* reported that Bush had announced that Zerhouni "would support government-funded work only on the approximately 60 cell lines developed before he made his decision, a compromise many scientists fear is too limiting."[94]

"This stand cost him some points with scientists," University of Pennsylvania ethicist Arthur Caplan believed.[97] Zerhouni would later join the scientific community in openly favoring support, an action that was described as "courageous."[98]

"Elias said 'Science should drive policy but not the reverse,'" according to Edward Miller, dean and CEO of Johns Hopkins Medicine and Zerhouni's chief at Hopkins.[8] "Basic sciences won't make advances in medicine alone so the institutes must work better together and be driven more by problems than by disciplines," both Miller and Zerhouni believe.[8] "Elias brings a business approach to the job. He's a leader in MRI imaging and has had experience in business—he helped to start a company," said Miller.[8]

National Institutes of Health Reform Act of 2006

On January 15, 2007, President Bush signed this omnibus reauthorization, the third in the history of the NIH. The NIH had not been reauthorized for 14 years, a delay caused, in part, by conflict over controversial issues such as stem cell research.*

The National Institutes of Health Reform Act increased the authority of the NIH director,[8,28,63,99] and Elias Zerhouni was the first director to benefit from the changes. Among its features, the Health Reform Act authorized the creation of:[100,101]

- A "Common Fund" for the NIH director to support specific research programs that he or she likes, apart from the funds provided for the institutes and centers. Previously, to fund

---

*The NIH can operate indefinitely under the provisions of its most recent reauthorization.

programs the directors favored, they had to personally persuade institute directors to provide the funds.
- A "Council of Councils" of 27 members selected from the Advisory Councils for each of the institutes and centers to advise on research proposals funded by the Common Fund.
- A "Division of Program Coordination, Planning, and Strategic Initiatives" (DPCPSI) in the director's office, which consolidates offices currently in the director's office, and a new Office of Portfolio Analysis and Strategic Initiatives that, among other responsibilities, administers and develops projects supported by the Common Fund.
- A "Scientific Management Review Board," which examines the use of NIH's organizational structure.

Although the Act specified that the number of institutes and centers may not exceed the current number of 27, it did not change the essential structure of the NIH and specifically did not consolidate the institutes and centers into smaller groups as Harold Varmus had recommended.[102] Nevertheless, Varmus said, he "considered the capping of the number of ICs to be a victory. It would have been nice to see some consolidation ... but I never expected my full-throated proposal to be adopted."[20]

Roadmap for Medical Research[4,5,11,25,37,39,46,54-59,65,72,75,76,85,94,100,103-125]

Soon after arriving at the NIH, Zerhouni started developing a strategic plan. "It was clear to me that science had changed, but that NIH had not. The convergence of concepts and methodologies applicable to all diseases made it necessary to break the traditional silos between institutes and disciplines," Zerhouni wrote.[126]

He sent a questionnaire to more than 300 nationally recognized leaders in academia, industry, government, and the public, that asked:

- What are today's scientific challenges?
- What are the roadblocks to progress?
- What do we need to do to overcome roadblocks?
- What can't be accomplished by any single institute but should be the responsibility of NIH as a whole?

This led to a series of meetings in Bethesda of institute directors, senior NIH administrators, and outside authorities to digest the information from the questionnaire and recommend solutions. Zerhouni stressed that he was not looking for "business as usual under another name." Instead, the groups should develop "exciting, enabling ideas and actions that can be clearly articulated to a wide audience."[127]

Then, in 2003, about 18 months after beginning the process,[115] Zerhouni rolled out the first version of his "Roadmap* for Medical Research." This particularly pleased clinical investigators who had felt underappreciated during the directorship of Harold Varmus, the quintessential basic medical science investigator. Zerhouni aimed to support innovative science and, in particular, clinical, interdisciplinary, and translational research.[125] He wanted to decrease the dominance of silos in the institutes and encourage more inter-institute collaboration.[13]†

"We're the National Institutes of Health, not the National Institutes of Basic Research," Zerhouni has been heard to say.[37] NIH intramural investigator Alan Schechter declares, "Elias's Roadmap is just what was needed."[25]

Constructing the Roadmap was "vastly time-consuming at many levels, involving many different parties across NIH," said Paul Sieving, director of National Eye Institute. "It was a difficult beast, but I'm enthusiastic about it."[123]

Zerhouni assigned a small part of the NIH budget to support Roadmap projects.[129] For the first three years, the institutes, most quite reluctantly,[37] provided the money—the pain of doing so exacerbated because the program was starting during a period of budgetary reductions.[110] In 2005, Roadmap's first year, the program received 0.8 percent of the NIH budget, in 2006 1.2 percent, and in 2007 1.64 percent or $465 million[79] of the NIH budget of $29.2 billion that year.

In the NIH reauthorization act of 2006, Congress created the "Common Fund" for the director to support Roadmap, which

---

* The word "Roadmap" was chosen because Zerhouni wanted the program to overcome roadblocks in the progress of clinical research.[103,125] Two winners of the Nobel Prize objected. "'Roadmap' indicates that you know where you're going, whereas science is not knowing where you're going," Michael Brown observed.[128] Arthur Kornberg, who won the Nobel Prize (1959) and fathered the winner of another (2006), wrote to me that he is critical "of 'Roadmaps' and similar buzzwords that demean the power of basic research."[116]

† Edward Miller of Johns Hopkins believes that the Roadmap "came out of our environment, flows from work Elias did here."[8]

Chapter Seven: Directors

became a congressionally mandated program. Subsequently, the institutes no longer had to contribute to the expenses of Roadmap,[125] though, as National Institute of Neurological Disorders and Stroke director Story Landis, comments, "The money still comes from one pool, and the dollars could have gone to the institutes."[85]

According to the legislation that established it, the Common Fund can increase to five percent of the total NIH budget, if the budget grows faster than the inflation rate.[100] Most of the money is directed into the extramural program.[114] Accordingly, each institute is expected to propose new Roadmap programs yearly.[85] Some expect that closing failing Roadmap projects will be difficult.[85] Roadmap will be subjected to an intensive scientific review when it reaches five percent of the budget.[100]

Since initially some of the funds for Roadmap came from programs that supported extramural research, complaints arrived from members of the extramural community,[46] "who were not given a vote on it," as one institute director complained. The years of flat budgets had increased tensions between extramurally supported scientists and the NIH leadership. "How dare you presume to tell us what to do or what it means to be a scientist?" is the gist of what they were saying.

"I don't agree with translational labs," said Gerald Weissmann, editor of the *Federation of American Societies for Experimental Biology* journal. "Individuals should apply for grants and get collaborators rather than using co-PIs [principal investigators]. You wind up subsidizing weak people to get program project grants funded. You don't need to direct science from the center."[130]

Within the NIH, not everyone liked the concept of the Roadmap, particularly in its first years, when the institutes were called on to support it with their budgets. "I was an agnostic about Roadmap when it started," said Richard Hodes, director of the National Institute on Aging. "Now I'm a real convert."[65]

The NIH Roadmap for Medical Research was designed to benefit, in part, the smaller or non–disease oriented institutes, like the National Institute of Environmental Health Sciences, and undertake projects that the institutes might not ordinarily support.[14] The budgetary limitations have made the NIH conservative about starting new programs,[123,131] and one of the purposes of Roadmap is to compensate for this tendency.

Roadmap funding is temporary. After a suitable time, investigators are expected to convert the source of the money that they receive from

Roadmap to one of the conventional funding mechanisms.[114] Since Roadmap was developed relatively recently, none of its programs have yet been fully evaluated.[57] "It's a work in progress," according to a senior scientist in the National Cancer Institute.[75]

Institutional Clinical and Translational Science Awards (CTSA)[11,52,57,59,71,98,106,107,114,125,128,132–139] are gradually replacing the General Clinical Research Centers (GCRC), a 60-year-old NIH-funded program for academic medical centers and research institutes. GCRCs support the cost of inpatient beds, ambulatory sites, professional and support personnel, and administration that clinical investigators need to perform their research. Zerhouni and his colleagues wanted to provide a more comprehensive infrastructure for clinical research than the older GCRCs supplied and remove obstacles they saw interfering with their performance, such as:

- Creation of fiefdoms by the GCRC directors who limited the research to their own interests
- Fragmented training programs
- Avoiding participation in multicenter trials
- Difficulty recruiting and retaining clinical and translational investigators. Translational research applies basic discoveries to clinical medicine. "Bench-to-bedside" is the frequently employed mantra.*
- Inadequate policing of safety by the institutional review boards (IRB) and increasing regulatory burdens
- Mounting overhead costs
- Inadequate sharing and coordination of the use of expensive facilities such as imaging and pathology
- Deficient informatics, the application of sophisticated computer technology and other storage and retrieval techniques to medical research and patient care.

The Clinical and Translational Science Awards are more inclusive and flexible than the GCRCs that they are replacing.† The

---

* Alan Schechter objects. "'Bench to bedside' is an oxymoron. 'Bedside-to-bench-to-bedside' is best. As a first approximation, I think that the NIH should divide its support for research about one third to clinical, one third to basic, and one third to translational."[25]

† The Clinical Center at the NIH can be thought of as a very large GCRC, a home for patients with rare diseases.[114,140]

investigators are expected to collaborate with investigators from different disciplines, including those in schools other than medicine.[125,135] Believing that the departmental structure typical of medical schools discourages collaboration, the NIH wants the CTSAs to contain fewer independent silos of disease-related groups.

As for funding, Zerhouni says, "Roadmap funds 40 percent basic research, 20 percent interdisciplinary and high risk basic research, and 40 percent clinical research thru the CTSA program. This is the same percentage of 60 percent basic/40 percent applied that I believe is the right combination for NIH and that I strived for."[141]

In the past, most grant applications came from one principal investigator, regardless of the extent to which colleagues contributed to the research. The NIH is encouraging two or even more investigators to apply as principal investigators on the same application since medical science is becoming increasingly complex, requiring investigators from different disciplines "to come to the table as equals,"[138] and to encourage team participation. Each PI in a multiple-PI application and grant is expected to be indistinguishable from an individual PI in a traditional application.[142] The multiple-PI model, however, may pose some challenges for institutions that significantly consider, in making promotion and tenure decisions, the ability of their faculty to compete for grants as a PI.

NIH no longer ranks schools and departments by the size of their budgets from the NIH. "Discontinuation of the practice preceded the implementation of the multiple-PI policy, but the anticipated difficulties in assigning credit under the multiple-PI system contributed to the rationale," said Ruiz Bravo.[142] Although dedicated rankers can derive their school's relative standing by computing data on the NIH's website, no longer will it be possible to simply look at the formerly available charts to determine where a school stands in relation to its peers.[143]

The CTSAs will not support strictly basic science research without clear translational implications, as was carried out in some of the older GCRCs, but will support biostatistics, bioinformatics, cores of highly technical equipment, and training.

The program will try to standardize among different institutions with CTSAs how the results of clinical research are reported, how the institutional review boards, which review grant applications for adherence to ethical standards for the care of patients and animals, operate, and how adverse events to patients are described.[125] The technical services must link with similar services at other CTSAs.[125]

Members of the CTSAs must take the Clinical Center's course on clinical research or a similar course developed at the CTSA.[144]

Among the 35 institutions that applied, twelve leading medical centers were chosen to receive the first grants in 2006.[135,145] A second group of 14 institutions was announced in June, 2008. The program will cost about $500 million per year by 2012, when about 60 CTSAs will be operating.[135] Many of the CTSAs will receive significantly more support than a typical GCRC used to receive.[106] Although part of the money for the CTSAs now comes from the director's Roadmap allocation, eventually all the support will be administered and funded by the NIH's National Center for Research Resources, to which the older General Clinical Research Centers have traditionally been assigned. A steering committee of investigators from the CTSAs meets regularly to provide advice to the NIH about the new program.[132]

The replacement of the GCRCs by CTSAs has not been enthusiastically received by those who must now develop the new administrative units.[107] Critics say:

- "The CTSAs are too big, too complicated, and overstructured."
- "Their scope is broad, perhaps too broad."
- "Our CTSA has less money to study patients than did our old GCRC."
- "People are in meetings all day."
- "They're a nightmare and will stress institutions that have them or are trying to get them."

David Nathan, from Harvard, no friend of the CTSAs, favors individual support for investigators rather than "grand plans with tons of meetings and mergers of departments that have no meaning. Small groups do the best science. What the NIH should have done was to have tougher site visits and get rid of the underfunctioning GCRCs, not create a whole new series."[98]

Some observers predict turf battles will develop between the different schools and departments incorporated in CTSAs. Despite these reservations, however, many leading academic medical centers have them or are applying for them. As the director of one of the new CTSAs said, "The only thing worse than having a CTSA is *not* having a CTSA."

Since the NIH specified that only one CTSA could be given to each degree-granting institution, Harvard was not one of the initial 12 because of difficulty melding the GCRCs at four of its teaching hospitals into one CTSA. Harvard tried unsuccessfully to persuade the NIH to change the rule. Since funding for each of the GCRCs will end in 2010, leaders at the medical school have developed a consolidated plan, which was submitted for the November, 2007, deadline. "Since everyone here has his own fiefdom, it's been incredibly stressful," said former associate dean for clinical programs, Daniel Singer. "People here have their day jobs to do in addition to trying to create a Harvard CTSA."[146]*

Garret FitzGerald, who led the successful application for the University of Pennsylvania, where he is chairman of the department of pharmacology, said, "It almost killed me!"[133] FitzGerald had fortuitously anticipated what was coming by establishing an Institute for Translational Medicine and Therapeutics before the CTSA concept had been circulated. Besides members of the school of medicine, Penn's application includes investigators from eight other schools within the university, plus the Children's Hospital of Philadelphia, the Wistar Institute, and the University of the Sciences in Philadelphia. "To bring it off, we had to resolve many intra-institutional problems, so we focused on the program first, the dollars later," FitzGerald said. "It was a huge physically and emotionally draining experience, an order of magnitude greater than writing SCOR† grants, but already it's helping to change the culture."[133]

The Penn program, which received $70 million from the NIH and $30 million from the participants, mostly the medical school, for the first five years, will be moving into a clinical and translational research facility attached to Penn's new ambulatory care building.

Appropriately, FitzGerald carries the title of Robert L. McNeil, Jr., Professor in Translational Medicine and Therapeutics at Penn. He is the only principal investigator of the first 12 institutions awarded CTSAs who is chairman of a basic science department. However, FitzGerald is a physician and was chairman of the department of medicine and experimental therapeutics at University College, Dublin, from 1991 to 1994.[133]‡

---

* Harvard got its CTSA in the spring of 2008.
† Specialized Center of Research.
‡ FitzGerald is the sage who told me, "Deans and chairmen of medicine exemplify a classic mismatch between aspirations and resources."[133]

The phasing out of the GCRCs brought complaints from governors and members of Congress—presumably alerted by distressed directors and medical school deans—that their states and districts might lose significant amounts of federal money.

Pioneer Awards[28,61,113,130,138,139] support extramural scientists who propose highly innovative studies in biomedical and behavioral research. The studies that potential pioneer awardees describe may be too novel, span too diverse a range of disciplines, or be at too early a stage to succeed in the traditional peer-review process.[72] Unlike most NIH grant mechanisms, these awards fund the person rather than the project. "What was needed," said Francis Collins, director of the National Human Genome Research Institute from 1993 to 2008, "are expensive high-risk projects, but it's not a free pass to craziness."[5]

The Pioneer Award program makes five to ten grants each year—too few, its supporters say[113]—and recipients must commit at least 51 percent of their research effort to the Pioneer project. The awards currently pay $500,000 in direct costs per year for five years and can be renewed once, after which the investigator must obtain support through the usual extramural system.[100,147]

New Innovator Awards are designed for scientists who received their last degrees no more than ten years ago. It funds 30 young scientists per year for direct costs of $350,000.

Molecular libraries,[28,139] a particularly popular innovation among the medical scientists, helps investigators in the public sector identify the small molecules—most medicines are in this class—that they need from a collection with a large-scale screening capacity.

Office of Portfolio Analysis and Strategic Initiatives (OPASI):[148] Elias Zerhouni established this new NIH office in 2007 to oversee Roadmap and "look across NIH's research portfolio and find better ways to manage it while respecting the turf of its 27 institutes and centers," as an article in *Science* magazine put it.[149] Zerhouni sees the new office as providing a permanent structure for Roadmap that will continue to function after Zerhouni's term as director ends.[150]

The first deputy director for the Office of Portfolio Analysis and Strategic Initiatives is Alan Krensky, a pediatric nephrologist and immunologist, who came to the NIH from Stanford University where he was, most recently, executive director of the Children's Health Initiative at the medical school and the Lucile Packard Children's Hospital.[100,150] Krensky's office opened with a budget of $3.3 million and a staff of 15, which in time is expected to grow to about

70 people,[150] leading at least one director to wonder if OPASI will become, in function if not in name, the twenty-eighth institute or center.[85]

Governance

Like the directors before him, Zerhouni has tried to curb some of the traditional independent authority of the institute directors exercised at the expense of the NIH director. To an extent he has succeeded in raising the power of the director's office relative to that of the institute directors, while helping the leaders and scientists in the institutes and centers "work together better."[113]

Before Zerhouni's reforms, the principal governing organization was the committee of directors of each of the 27 institutes and centers.* Since everyone, in effect, had a veto on the director's proposals, decisions were difficult to make and institute. The committee of institute and center directors continues to meet to discuss scientific, but not administrative or financial, issues.

Recognizing that a better system was needed to facilitate the analysis of problems and management of decisions, Zerhouni created a steering committee of ten institute directors as his senior advisory group. The directors of the three largest—National Cancer Institute; National Heart, Lung, and Blood Institute; and National Institute of Allergy and Infectious Diseases—are permanent members. The directors of seven other institutes serve for staggered, three-year terms.[57] Zerhouni usually chairs the twice-per-month meetings.

Members of the steering committee chair each of six working groups on: Facilities, Management and Budget, Intramural Program, Extramural Program, Information Technology, and the Roadmap. These groups, which meet at least monthly, report their deliberations to the steering committee and Zerhouni for decisions. Despite his having created the steering committee, however, Zerhouni has been known to occasionally make independent decisions outside the committee.

In keeping with his engineering and managerial background, Zerhouni has steadily brought greater control and authority to the director's office. This contrasts with the style of Harold Varmus, who

---

* As noted before, the institutes and centers as a group are called "I/Cs" in NIH-speak.

was more comfortable with authority's resting with the individual investigator, the model he had grown up with.[109]

Background

Elias Zerhouni, the first NIH director not to have an American medical degree, was born in Nedroma, Algeria, and graduated in 1975 from the University of Algeria School of Medicine. Zerhouni came to the Johns Hopkins Hospital for a residency in diagnostic radiology and in 1979 was appointed assistant professor of radiology in the Hopkins medical school. He spent 1981 to 1985 at the Eastern Virginia Medical School in Norfolk and then returned to Hopkins, where he was promoted to full professor in 1992 and named chairman of the department in 1996. Zerhouni later became vice-dean for research and for clinical affairs as well as executive vice-dean at the medical school, where, according to his colleague Guy McKhann, "Elias was extraordinarily well respected at Hopkins. A real straight-shooter."[72]*

Like Bernadine Healy, who had worked in the Office of Technology Policy while Ronald Reagan was president, Zerhouni served as a medical consultant to the president. In 1991, he took a leave of absence from Hopkins to help found Advanced Medical Imaging Institute in Norfolk. His research has centered on computerized axial tomography (CAT scanning) and magnetic resonance imaging (MRI). He has authored or co-authored 157 peer-reviewed publications and holds eight patents.

Zerhouni had never worked at the NIH before becoming director as had most of his predecessors. Directors since 1975 had come to the NIH early in their careers for advanced training to become biomedical investigators and then had distinguished careers as faculty members at leading medical schools or as members of the intramural program at the NIH.† Zerhouni did, however, serve on the National Cancer Institute's Board of Scientific Advisors from 1998 to 2002.[151,152]‡

---

* Zerhouni commutes from Severna Park, a Baltimore suburb, where he built a house in a style reminiscent of his Algerian origins.[8]

† Donald Fredrickson (director from 1975 to 1981) at the NIH, James Wyngaarden (1982–1989) at Duke, Bernadine Healy (1991–1993) at Johns Hopkins, and Harold Varmus (1993–1999) at the University of California–San Francisco.

‡ See http://www.nih.gov/about/director/directorbio.htm and file:///k:/NIH%20CVs/Zerhouni.htm. for Zerhouni's biographical sketch and CV.

Opinions[5-7,11,13,26,29,31,33,37,39,46,49,53,54,57-61,63,65,71,72,75,76,83, 87,96,97, 103-106,113,118,128,132,139,153-155]

Colleagues think that Zerhouni's coming from another country makes him a better chief. "He hasn't grown up with 'That's the way we do things.'" He is seen as articulate, courageous, forthright, hard-working, visionary, "incredibly smart," a quick study, supportive of his colleagues, understanding of organizational issues, "good at motivating people to work as a team," and "willing to compromise," making him "particularly adept at building consensus." He's analytical and process-oriented "like an engineer," thinks in quantitative terms, and engages in "systematic, rather than linear thinking." He is regarded as "almost hyperrational, always badgering people for more facts," and a leader who "sees the big picture." Most think his business experiences have helped the NIH and praise his work in breaking down administrative barriers, particularly between the institutes.

Josephine Briggs, director of the National Center for Complementary and Alternative Medicine, regrets that Zerhouni did not tackle the long-standing NIH policy that intramural clinical researchers may not collaborate with extramural investigators in multi-institutional clinical trials. She notes that "study sections are also forbidden to look at intramural programs, which handicaps the work of intramural clinical investigators."[106]

Science journalist Barbara Culliton finds Zerhouni "different from previous directors in coming from applied technology. He's more informed about bioinformatics and how technology can help move science forward. He's very contemporary in that sense."[37] One of his colleagues, while praising Zerhouni's appreciation of good science, said, "He doesn't see where science is going as well as Varmus can."

His colleagues admire Zerhouni's ability to deal with Congress— "he speaks to both sides of the aisle"—where he is seen as effective, diplomatic, and adept at dealing with issues like stem cells and abortion, and has "the backbone to face down the politicians." Former U.S. representative Paul Rogers, a longtime and ardent NIH supporter, calls Zerhouni "excellent. He's bringing the institutes together."[153] His listeners appreciated the thoughtful talk that Zerhouni, who is a Moslem, gave on an anniversary of the September 11th attacks.

Zerhouni has been able to maneuver skillfully as a political appointee in an ideological administration with little apparent desire to support NIH and in a time of strong political partisanship. He found himself "in an extremely tough job trying to please both politicians and scientists."

Whereas some complain that he's not the basic scientist that Varmus was, others recognize that Zerhouni, the applied scientist, has a sophisticated understanding and appreciation of basic science. He is recognized for favoring entrepreneurial, public–private partnerships like Bernadine Healy did, but "it's a slippery slope working with Big Pharma," as one former NIH staff member warned.

Roy Vagelos, the former CEO of Merck and Co., is "bothered by the recent tendency of the NIH to support drug discovery at universities." This work, he believes, is best done by pharmaceutical companies, which have the staff and facilities to carry out this complicated work. "For the most part, university people are not set up to make new drugs," he said. "Universities, despite the financial advantages of acquiring patents of commercially successful drugs developed in their labs, should emphasize discovering new basic information and training the scientists of the future."[156]

Although one critic sees Zerhouni as "seeking the trendy," and another avers that Roadmap has not been a "public relations success," most praise him for developing interdisciplinary programs and improving communication between the institutes. "He's looking for bold initiatives that will be transforming," said Keith Yamamoto of the University of California–San Francisco, who is co-chairman of the external committee reviewing the peer-review system.[155]

Many people find Zerhouni visionary, but pragmatic, and the right leader for the time. "Elias is trying things that the bureaucracy can't stand," comments Craig Venter, a human genome investigator who worked at the NIH from 1982 to 1992 before leaving to form his own research institute. "The disease orientation of the institutes is a constant problem. It's hard to work across them, unlike where I work now [in the J. Craig Venter Institute] where I can put together multiple interdisciplinary teams. The NIH is not a risk-award system."[61]

Zerhouni Leaves

On September 24, 2008, Zerhouni announced that he would leave the NIH as director at the end of October. He told reporters that he planned "to write about his time at the health agencies before accepting

Chapter Seven: Directors   191

another job."* Zerhouni decided to leave the agency before the election, "so there is a clear sense that whoever wins the election, NIH has to be a clear priority in their mind."[157]† Raynard Kington, the deputy director, would serve as the NIH interim director "for the remainder of the Bush administration."[157]‡

On October 30, 2008, Tony Fauci read a tribute that praised Zerhouni as a highly successful NIH director and included the following:[159]

> As an individual you exude a sincere warmth and caring for your friends and colleagues. You are kind and compassionate, but you are tough when you have to be.... You are a great listener and have extraordinary patience in hearing people out, always accumulating, synthesizing and analyzing information.... I can say with confidence that I speak for the IC Directors in saying that you will be greatly missed.

---

* In response to my inquiry about his post-NIH activities, Zerhouni wrote on July 16, 2009:[92]
"At the request of Ed [Miller, dean/CEO of Johns Hopkins Medicine] I agreed to serve as an unpaid senior adviser to Johns Hopkins Medicine and returned as professor on the faculty of the departments of radiology and biomedical engineering. I am also advising the Gates Foundation and have been elected to the boards of the Lasker Foundation, Research!America, and the Mayo clinic as well as the board of the King Abdullah University of science and technology.... I have been appointed as the chief scientific adviser to a sister publication to *Science,* which will launch this fall, entitled *Science: Translational Medicine....* I am pleased with that as I think that our field is hurting from the gap that has developed between basic and clinician scientists.
"As far as career is concerned I have decided to take a full year before deciding what I want to do next and refused to consider any full-time job until then. Nonetheless I remain very active and involved in many venues."

† See http://www.nih.gov/news/health/sep2008/od-24.htm for the NIH's announcement.

‡ On July 8, 2009, President Obama nominated Francis Collins, former director of the National Human Genome Research Institute (see above), to succeed Elias Zerhouni as director of the National Institutes of Health.[158] The Senate confirmed the appointment on August 7, 2009.[158a] The author of an op-ed contribution in *The New York Times* objected to the appointment of Collins, who is deeply religious. "Must we really entrust the future of biomedical research in the United States to a man who sincerely believes that a scientific understanding of human nature is impossible?"[158b]

CHAPTER EIGHT

Controversies

Conflicts of Interest in the Intramural Program

The rules

BY THE 1990s, the pharmaceutical companies—"Big Pharma," as the group was becoming known—and the biomedical firms were offering increasingly large financial inducements for investigators of NIH-quality to work with them.[1] Such consulting for pay had long been forbidden at the NIH.

When Harold Varmus became director, he loosened these restrictions, a decision praised by most within the NIH and many outside.[2] The government rules—Varmus had not worked in government except briefly when a fellow at the NIH during training—did not apply at the University of California–San Francisco where he had spent most of his career. Varmus believed that some of the rules were inhibiting his ability to recruit and retain scientists at the NIH. "The most frequent non-scientific topics the intramural people talk about are the conflict-of-interest rules and the uncompetitive salaries, particularly for senior scientists," said one of the institute directors.[3]

Varmus was not alone in believing that outside interests for NIH investigators help advance science and public health and develop useful contacts.[1,2,4-19] The Varmus rules allowed, among other provisions:[20,21]

- Senior staff, with the exception of presidential appointees, to conduct outside activities without the severe limits previously imposed.
- Employees to receive, without limit, money, stocks, or stock options for activities outside the NIH and to spend time, again without limit, on activities that do not interfere with their work at the NIH.
- Employees to receive compensation for speaking or lecturing to outside organizations.
- Amount and type of compensation did not need to be disclosed to the NIH.

Varmus's rules significantly liberalized what compensation intramural scientists could receive from external work. Although the effect of this change was to reduce oversight and increase permissiveness,[12,22] NIH policy required that employees receive authorization from their supervisors to perform consultations for pay and report annually all external activities they were paid for,[14] but no formal peer review through a formal conflict-of-interest committee, as practiced at many institutions, was implemented.[21]

A few[3] at the NIH bent the rules—"ran with it and were caught," as former director Bernadine Healy described what happened[23]—and the effects were traumatic. Most would be criticized for not requesting permission for, or not reporting, the outside work they were doing. In the "open society" of the NIH, the amount of oversight exercised in the institutes and centers about such activities was minimal.[22] Others came under scrutiny because of the large amounts of money they were receiving.[24] "Some people just do dumb things," said Barbara Culliton.[2]

Probably the most egregious example was Trey Sunderland, who became what the medical ethicist Arthur Caplan called "the poster child of what was wrong."[1] It was "a dark chapter in the history of the NIH," one of the institute directors said.[25]

The case of Trey Sunderland

Three days before Christmas in 2006, a federal judge ordered Pearson Trey Sunderland III, the former chief of the geriatric psychiatry branch at the National Institute of Mental Health, to pay $300,000 for illicit compensation that he had received from Pfizer, Inc., the world's largest pharmaceutical company.* The court allowed Sunderland 18 months to make the payment and charged him no interest. He was given two years of supervised probation and assigned to perform 400 hours of community service. The sentencing followed Sunderland's guilty plea on December 8th to a misdemeanor charge of violating conflict-of-interest rules.[26,27]†

His offense was not informing his superiors at the NIH about the Pfizer payments as regulations required.[27,30,31] "It is illegal," a U.S. attorney was quoted in the *Los Angeles Times*, "for any federal employee to make an official decision that directly affects their financial interest, unless they ... get approval from the government."[27]

Pfizer had paid Sunderland from 1998 to 2003 while he was collaborating with the company in his official capacity as an investigator of Alzheimer's disease. The problem came to light when staff members of a congressman compared records of what Sunderland had received from Pfizer and what he had reported to the NIH.[25]

Sunderland's group had sent samples of spinal fluid taken from NIH patients to Pfizer for the company's scientists to study for clues that might lead to treatments for the disease. Sunderland had received NIH approval for this work. However, Willman reported, "NIH officials have said that if Sunderland had asked to moonlight for Pfizer, the request probably would have been denied because the activity overlapped with his official duties."[27]

Trey Sunderland, as he is known (and described as "charming beyond belief")[11] came to the NIH with a B.A. in psychology from Harvard College, an M.D. from George Washington University, and postdoctoral training in medicine and psychiatry at Harvard-affiliated hospitals in Boston. He is widely published in the psychiatric literature emphasizing aging and Alzheimer's disease and served as chairman of the NIMH institutional review board for ten years. Among his awards

---

* See http://www.usdoj.gov/usao/md for the Justice Department report. Click on Press Releases, then 2006 Press Releases, then 12/22/06: "NIH Senior Scientist Pearson Sunderland Sentenced on Conflict of Interest Charge."

† Much of the information in these paragraphs has been derived from the reporting of David Willman of the *Los Angeles Times*[28,29] and Rick Weiss of *The Washington Post*.

have been, ironically, the NIH Director's Award in 1998 and the National Institute of Mental Health's Exemplary Psychiatrist Award in 2000.[32] "We saw him as a very good intramural investigator and an articulate spokesman," said Thomas Insel, current director of the institute where Sunderland worked.[33]

As he was a member of the Commissioned Corps, the NIH did not have the authority to fire Sunderland, and he remained a government employee,[24] to the anger of members of Congress.[34] Finally, by May 1, 2007, with all outstanding issues settled, he left the NIH.[12]

Degree of conflict of interest

Although Sunderland was the most publicized case, other well-known investigators were caught in the web that developed with the loosening of the rules.[36,37] Richard Klausner, a political appointee[21] and director of the National Cancer Institute, had accepted an award from the University of Pittsburgh. The problem was that Pittsburgh had grants from Klausner's institute. Was the university honoring him to thank him for the money and ask for more? Investigation revealed that the award to Klausner was *bona fide*, that there was no apparent connection between the award to Klausner and the NIH grant, and the issue went no further.[12]

James Battey, the director of the Institute on Deafness and Other Communication Disorders and a 24-year employee of the NIH, thought that the new rules might force him to resign.[38] Battey manages a trust for his family, and having to sell the stocks in the pharmaceutical companies would produce sizable capital gains, with accompanying taxes. After involved discussions, Zerhouni and his staff decided not to force Battey to sell the interdicted securities in the trust and ruled that he did not have to report these holdings. Battey explained, "It's actually an issue of perception rather than reality. Nothing I could do in my professional capacity could affect the stocks in Big Pharma. Might be different if we were dealing with a small company."[25]

Blue-ribbon Panel

As the stories in the press about conflicts of interest spread, in January, 2004, Zerhouni formed "a blue-ribbon panel" to investigate the problem.[39] The co-chairs were Bruce Alberts, president of the National Academy of Sciences, and Norman Augustine, chairman of the executive committee of Lockheed Martin.[40] With regard to conflicts of interest, their report, which was issued on June 22, 2004,[41,42] advised:

- "NIH senior management and NIH intramural employees ... should not engage in consulting activities with pharmaceutical or biotechnology companies.
- "Intramural scientists conducting research with human subjects ... should not be allowed to have any financial interest in or relationship with any company whose interests could be affected by their research or clinical trial.
- "Employees must disclose all relevant outside relationships and financial holdings.
- "NIH scientists should be allowed to receive compensation for teaching, speaking, or writing about their research [and] be allowed to engage in compensated speaking, teaching, and writing for professional societies and for academic and research institutions as an outside activity providing that all ethics review and approval requirements are met.
- "NIH should ... permit employees to be identified by their title or position (and institutional affiliation) when engaged in teaching, speaking, or writing as an approved outside activity."

Congress was dissatisfied with the report, a preliminary version of which they received in May of 2004.[43] Traditional congressional supporters of the NIH there felt that their confidence had been undermined, that some members of the NIH were serving their own interests rather than the public good.[44] James Greenwood (R-Pa.), chairman of the Energy and Commerce Subcommittee on Oversight and Investigations, and other committee members assailed Zerhouni and the co-chairs of the panel with what the reporter for *The Washington Post* characterized as "four hours of often withering criticism."[45] Joe Barton (R-Tex.), chairman of the Energy and Commerce Committee, said, "We have found NIH to be less than cooperative, and that's going to change. They can cooperate cooperatively, or we will make them cooperate coercively." Greenwood was especially critical of the legal staff at the Department of Health and Human Services, whom he accused of "delays and obstinacy."[45]*

New Rules

On February 3, 2005, the independent office of government ethics and the Department of Health and Human Services announced stringent new interim rules for the employees of all its agencies, including the National Institutes of Health, about being paid to consult for, or hold stock in, pharmaceutical firms.[48] When Zerhouni described these interim rules at an open meeting, most of those present attacked him and the new proposed regulations, and few gave support. NIH employees at the meeting said, according to *The Washington Post*:[21,49]†

- "If we really want to reassure the public . . . why don't we apply these to everyone who gets an NIH grant?"
- "Does this apply to the Department of Energy? To the Department of Agriculture? To the Defense Department?"
- "Even my secretary is going to have to sell her stock. How much sense does that make?"

"Most irritating" to the NIH people, the reporter felt, "is the rule that will require thousands of employees—and their spouses and dependents—to divest themselves of all stock holdings in drug, biotech and other medically oriented companies. Even lower-ranking employees with no influence on grants or policies will be limited to individual holdings of $15,000. All were required to make those divestitures within 90 days, at a time, as one speaker put it, that much of that industry 'is at the bottom of a cycle.'"

---

* Varmus believes that Zerhouni has had to deal with a Congress that is more hostile to the NIH than when he was director.[46] Phillip Gorden, the former director of NIDDK, agrees. "Interaction between the Department [of Health and Human Services] and Congress was very effective during Varmus's time. Now we're in this macro-political environment in which little is going to happen."[47]
† "We always have to deal with '*The Washington Post* syndrome'" observed a longtime NIH scientist. "No one likes to see his name in the paper in such a situation."[50]

The new interim regulations that Zerhouni announced specified that NIH employees:[12,21,51-57]*

- May not consult with, or speak at, institutions that receive NIH funding or with pharmaceutical and biotechnology companies, and health-care providers and insurers.
- May, by themselves and together with their immediate families, own no more than $15,000 in the securities of any one pharmaceutical, medical equipment, or biotech company and no more than $50,000 in all sector mutual funds that specialize in such companies, if the employee is a senior-level employee.

The theory behind such restrictions on outside activity is that the United States government pays NIH employees to do their work and expects them to devote their efforts full-time to their jobs. Consequently, NIH scientists should not consult with certain biotechnology and pharmaceutical companies, collect honoraria for giving talks at universities and professional meetings, receive fees from serving on editorial boards, or start new companies.[6,25,46]

Among the rules' peculiarities:

- A cancer researcher was told that she could not accept a $200 train ticket from a physician's education group to present a paper at a conference in New York.[38]
- Another scientist was told that he could not accept an unpaid adjunct professorship at Johns Hopkins because, since the appointment came with free parking, he might be unduly influenced in favor of the university.[59]
- For an NIH physician to be appointed as an unpaid lecturer at Harvard, the NIH was required to review the university—from which the doctor had received his undergraduate degree—for compliance with NIH ethics policies.
- NIH investigators invited to participate in a conference in Baltimore could not charge the government for staying in a hotel there since Baltimore was less than 50 miles from the NIH,[60] and they had to

---

*The NIH retained an independent research organization to survey the reactions of its employees to the new rules as they might affect recruitment and retention. 8,000, or about 48% of the staff, responded. See reference[58] for details.

commute daily from their homes in the Washington area.* (This is a federal rule applying throughout the government.)[21]
- NIH investigators who receive honoraria by mistake can't contribute the money to a charity that supports research on the disease they are studying. It must be returned, sometimes with troublesome tax complications.[6]

Congress questioned the receipt of payments by numerous NIH investigators who appeared to have engaged in outside activities with pharmaceutical companies without the requisite approval. An investigation revealed that a significant number of the accusations were caused by bookkeeping discrepancies or misidentification of the party involved in the activity.[59] When the review was completed in July, 2005, the NIH reported that, of the 103 employees investigated, 44 had violated conflict-of-interest rules, 37 had not, and 22 were still being reviewed.[61,62] Of those who had failed to request authorization for their work, "less than a handful would have been denied permission," speculated Holli Beckerman Jaffe, the director of the NIH Ethics Office, "because at the time the approval process, as applied to the questionable activities, was more lax."[24]

This news aggravated the unhappiness at the NIH, where members said that Zerhouni should have waited for these facts before succumbing to pressure from Congress and his bosses at the Department of Health and Human Services.[59] Senior NIH investigator Abner Notkins said, "All of us are in favor of strong regulations to avoid conflict of interest, but the new rules go to an extreme." Notkins heard that some NIH investigators were thinking of leaving and others who were being recruited deciding not to move because of the new rules.[14] At the peak of the controversy, "I heard people say," remembers Zerhouni, "'if it feels good, it's unethical.'"[63]

The director found himself arguing with the Department of Health and Human Services that the restrictions would exacerbate problems with recruiting and retaining scientists,[14] just the rationale that Varmus had previously used to justify relaxing them. His colleagues at the NIH joined Zerhouni in opposing the rules that DHHS was imposing on the

---

* Ezekiel Emanuel, chairman of the department of clinical bioethics in the Clinical Center, remembers being told that one researcher who was forced to commute was killed in a traffic accident.[8]

director. The new factor was that the Congress that was now pressuring the administration to repair the practices had given rise to the troubles. Leaders at the NIH heard of threats that Congress would reduce the budget. "It was a very serious business," said one NIH administrator; "not a joke."

Zerhouni inherited problems that had occurred before his directorship began.[50,54] He let it be known that he felt that those who had been caught had "shot me in the back by not disclosing their activities in time for the blue-ribbon panel to address" them, and that Varmus had safeguarded some of those Zerhouni had to remove. "Wink, wink. Don't tell me," was how Zerhouni described the atmosphere in Varmus's time.

Others said:

- "A buzz saw hit him."[64]
- "He really felt betrayed, putting him and the institute at risk."
- "Elias needed to put out the fire. He had to regain the faith in the NIH of the Congress and the public."[9]
- "The rumor is that lawyers at DHHS forced it down Elias's throat."
- "Some thought that the director could have fought DHHS more strongly."[65]
- "I blame Congress for jumping on the NIH."

"Zerhouni's original policy made no sense," said medical ethicist Arthur Caplan. "He came down so hard that there was no room to negotiate."[1] Many at the NIH felt that Zerhouni went too far, that the rules were "draconian,"[65] and that he had placed the NIH "in a fishbowl."[66]

"It was an overreaction," said Kenneth Shine, former president of the Institute of Medicine.[67] Even *The Washington Post*, which had zealously reported the NIH's troubles, editorialized that the rules were too strict.[68] Caplan believes,[1] and Zerhouni later acknowledged, that his initial response was "part of a strategy, of taking the high road from Congress and those in the administration in favor of draconian rules" that he would then use to later adjust and modify the proposed final rules.[21]

Vincent DeVita, a former director of the National Cancer Institute, believed that the first set of regulations, announced after Zerhouni was "sandbagged" by a few employees who disregarded the rules, "shattered morale in the intramural program. He should have fired the bad ones and not punished all the others. The rules were draconian to the wrong people."[7] NIH ethicist Ezekiel Emanuel lamented that "scientists are not politically savvy, are wary about using pressure such as media, and are not the most articulate people in non-science issues."[8]

"If you work there," said David Nathan from Harvard, who had trained and worked at the NIH, "you have to take the veil."[69] With the restrictions on travel, it's become "Fortress NIH," said Gerald Fischbach. "There are philosophers walking there."[66]

"The NIH has become more politicized," declared Zach Hall, former director of the National Institute of Neurological Disorders and Stroke. "Political control through the DHHS has tightened. It became a good time to recruit from the NIH."[70]

Zerhouni Revises the Rules

When members of Congress learned about the new interim rules, some concluded, like many at the NIH, that Zerhouni, though well intentioned, had exceeded what was reasonable, and they requested that he delay instituting the changes. "A few others understood that I was deliberately turning the onus on Congress and the Administration to finalize the rules," Zerhouni said later, "rather than NIH being seen as too lenient and thus vulnerable to attack as loose on ethics."[21] Senator Tom Harkin, one of the most reliable supporters of the NIH on Capitol Hill, said, "sometimes we tend to see a conflict of interest, and we go overboard."[71] Health and Human Services Secretary Michael Levitt decided to delay application of the rules until July 3rd.

The executive committee of the Assembly of Scientists,[72] made up of all NIH investigators who wish to participate, met frequently with Zerhouni "to mellow [sic] the restrictions," remembers Abner Notkins. "Zeke [Emanuel, the resident ethicist] had endless questions and suggestions."[14] The director also had to respond to more than 1,700 letters with strongly worded complaints from colleagues,[14] comments from Congress,[73] and a survey showing that "nearly 40 per cent of the scientists conducting hands-on research at the National Institutes of Health say they are looking for other jobs."[74]

The NIH gradually "clarified"[60] the rules. By the spring of 2007, in accordance with revised regulations from the Department of Health and Human Services, investigators were allowed to:[52,60,75,76]

- Perform professional work when not on duty at the NIH.*
- Participate in, and receive honoraria for, editing or writing textbooks, lecturing in continuing education courses, or speaking as recognized experts. However, when talking about their new work, investigators may not receive honoraria, since making discoveries is what the government is paying them to do.
- Consult without honoraria but with payment of expenses for engagements of a general academic nature including with industry. NIH scientists may not, however, consult on their primary research interest, unless the NIH is officially collaborating with the company on a research protocol.

Investigators may not:

- Own more than $15,000 of stock in a company supplying a drug they are studying, or more than $50,000 in a sector mutual fund that might include shares from the company. "Sell it, or do not participate in the study," they are told. The rules for senior-level employees are even more stringent. These include at the institutes: directors, deputy directors, clinical directors, and most senior directors responsible for the extramural programs.[12,22]†
- As authors or editors of books identify themselves as from the NIH, but must instead use "Bethesda, Maryland," unless the work has been performed as an official government activity.
- Hold leadership positions in professional medical and scientific organizations.

In an editorial, *The Washington Post* explained:[78]

> In the end the National Institutes of Health got it about right. This year, NIH proposed tightening its ethics guidelines to such an extent that the rules would have unnecessarily penalized

---

* Widely known as "moonlighting," in which physicians, particularly trainees, supplement their income by working at hospitals when they were not required for duty at the NIH.[60,75]

† "Those of us who can influence where the money goes have to be particularly careful," said Stephen James, a division director of the extramural program at the NIDDK. "We can receive travel expenses but no honoraria and no consulting fees. None of us may talk about our research until it's published so that the place where we talk doesn't get any advantage. Some of us only give review talks."[77]

employees and ultimately harmed the ability of the institutes to attract and retain top scientists. In issuing the final version of the rules Thursday, the agency backed away from its most draconian proposals while putting in place needed restrictions on outside consulting.

The issues that the conflict-of-interest brouhaha has raised, however, persist. "There's an impression," said Arnold Relman, former editor-in-chief of the *New England Journal of Medicine*, "that there's much internal resistance and ambiguity" about how vigorously the NIH is regulating conflict of interest among its employees. "How carefully are the policies followed and instituted?" he asked.[79]

"Conflict-of-interest issues have sustained wide swings of the pendulum," said Philip Pizzo, the Stanford dean who worked at the NIH for 23 years and was a member of the blue-ribbon panel. "What was permissible in part of the 1980s, then became not permissible until the mid 1990s, and was severely curtailed in 2005."[44] The more recent changes in policy, from draconian* to somewhat more liberal, reflect the pressure that the recently appointed Zerhouni had to respond to.[80] "Zerhouni did the best he could with Congress breathing down his neck and $28 billion at risk," said a senior investigator.[81]

Why Did the NIH Get into this Trouble?

One observer believes that leading reasons are the size of the institution and the belief of many of the scientists that they are independent entrepreneurs rather than public servants.[11] As for progress, one branch chief said, "We're recovered from the nadir of the problem. A lot of it was pretty silly and counterproductive."[82] "It was overdone," believes Gerald Fischbach. "Trey [Sunderland] was the exception."[66] Others agreed that few took advantage of the consulting rules.[83]

As for Zerhouni's role, "despite the conflict-of-interest turmoil, which was not of his making," Carl Roth, a longtime NIH employee, said, "he defended the NIH well while mitigating the damage and negotiated favorable 'terms of surrender'."[50]

---

* The word "draconian" seems to have been specifically invented for the conflict-of-interest controversy at the NIH. Many people used it and newspapers printed it when discussing the topic.

Nevertheless, "the 'period of unpleasantness' that started in July, 2003, continues," said Jaffe.[12] To deal with the continuing conflicts, in classic bureaucratic fashion, the number of positions in her office has grown from two people in October 2003 to 21 in 2007. "The informed community, as well as those of us in the NIH, have to understand that the NIH is four-plus against corruption and collusion with the pharmaceutical and biomedical companies," added Milton Corn of the National Library of Medicine.[80]

"The real risk is that NIH research could become targeted like drug company research," said Marcia Angell, author of a cautionary book about drug companies.[84] "Research should be done where it leads, not where dollars might be made. This could corrupt the NIH, which is such a great institution."[85]

"Public confidence in the NIH remains of critical importance," declared Kenneth Shine, who was president of the Institute of Medicine when its 2003 report on the NIH was being written. "The consultation issue was very embarrassing to it. The NIH has to be pristine,"[67] and as Robert Wiltrout from the cancer institute added, "beyond reproach."[19] Michael Gottesman concluded, "This is, by far, the major driving force for policy. NIH credibility is paramount, and Dr. Zerhouni knew this immediately."[60]

Conflicts of Interest in the Extramural Program

In the June 8, 2008, issue of *The New York Times* a story appeared about three psychiatrists at the Harvard Medical School who allegedly did not report the full amount of money they received for consulting to the university or, in two cases, to the Massachusetts General Hospital, which was administering their grants.[86] In October, *The Times* reported that an Emory University psychiatrist failed to inform his university about receiving at least $1.2 million from pharmaceutical companies.[87]

The episodes were uncovered by Senator Charles Grassley (R-Iowa) whose staff found that the amount of money reported by the doctors was less than the amount the pharmaceutical houses had paid them.[88] Grassley has found similar problems—"detonations," as *Science* writer Jocelyn Kaiser described the revelations—at other universities.[89] She wrote, "Grassley, who has accused NIH Director Elias Zerhouni of 'lax' oversight of extramural research, wants to require drug companies to report all payments to physicians in a public database."[90]

Chapter Eight: Controversies

With respect to conflicts of this type in the extramural program, the NIH expects faculty members to accurately inform their institutions about funds paid to them by organizations outside their primary place of employment. When the rules governing such reporting are not followed, "we work with the institutions since they are ultimately responsible for administering the grants and directly managing conflict-of-interest issues," explained Norka Ruiz Bravo, deputy director for extramural research. "Possible NIH actions range from requesting information about the institution's financial conflict-of-interest procedures and reviewing the institutions' actions in a particular case, to suspending funding for the award until the matter is resolved."[91]

Stem Cells

Throughout its more recent past, the NIH has had to respond to political constituencies that have interfered with the freedom of NIH-supported investigators to work where their spirit and imagination might take them. In the 1970s and 1980s, the issues were research on fetal tissues [10,16,47]* and recombinant DNA;[16,92,93]† in the 1990s, cloning; and more recently, stem cells.

Stem cells‡ are pluripotent cells that can become any of the specialized cells that complex biological systems like humans need. Thus, under proper conditions, a stem cell can develop into a muscle cell, a skin cell, a brain cell, etc. This extraordinary property has stimulated much interest among medical investigators, who believe that stem cells may be applied for therapeutic purposes.

Stem cells can be obtained from several sources: human embryos, the source of much of the controversy; certain cells in adults and children;

---

* Because the source of the tissue may be aborted fetuses, federal funding for research was restricted in the administrations of Ronald Reagan and George H. W. Bush.
† "Recombinant DNA" refers to nuclear material that has been modified by nuclear material introduced from another organism. The anxiety was based upon concern that, for example, a virus known to cause cancer in animals might combine with an otherwise innocuous bacterium and become established in humans, rendering the recipient at risk for cancer. Investigators initiated the concern about this procedure, which was then adopted by politicians and other individuals and groups.
‡ A cell (from *cellula*, Latin for "a small room") is the smallest self-contained biological unit. In addition to carrying out the functions necessary to live, a cell can reproduce itself. Some organisms, such as bacteria, have only one cell, an ostrich egg being the largest known cell. Humans have as many as 100 trillion cells.

blood in the umbilical cord; and amniotic fluid. Stem cells are now used in the treatment of some forms of leukemia and other disorders of the bone marrow, but, as of the spring of 2008, their usefulness in other diseases has not been demonstrated.

Embryonic* stem cells are taken from embryos generated during *in vitro* (outside the body) fertilization. In this process, eggs are removed from women unable to conceive normally. The eggs are sustained artificially, fertilized with sperm, allowed to grow for a few days, and then implanted in the woman's uterus. More eggs are usually harvested and fertilized than are needed for implantation. Those not used for deriving stem cells are frozen for later use or destroyed.

Embryonic stem cells are harvested from embryos in an early stage of development—in humans, about four to five days after fertilization—by a highly sophisticated technique first reported in 1998.[94] Removing the stem cells prevents the cell from progressing further, in effect killing the embryo. Those who oppose the use of embryonic stem cells base their objection on the destruction of a potential human life in the process of obtaining the stem cells.

Politics and Stem Cell Research

On August 9, 2001, President George W. Bush, Elias Zerhouni explains, "for the first time permitted federal funding of stem cell research but strictly limited it to the cell lines then in existence, which were thought to number 78 but ultimately would number 23 and would prove insufficient to support the science."[21]†

Investigators could now be supported for work with stem cell lines in existence before that date as long as the cells were no longer needed for reproductive purposes, the investigators had been provided with informed consent of the donor and there was no financial incentive for the donation. This policy prevented any research on newly derived stem cell lines, except as specified by investigators in the intramural program—each being an employee of the federal government—and

---

* An embryo is a fertilized egg that has reached an early stage of development.
† Jay Lefkowitz writes in *Commentary* (January, 2008): "The next day, at Yale, he [George W. Bush] sat down with Harold Varmus."[95] Varmus recalls, "I never 'sat down' with Bush or discussed anything of substance with him.... When I met Bush in a receiving line at Yale in 2001 and said the phrase 'stem cells', his eyes grew wide and he yelled to Andy Card (his chief of staff): 'Andy, come on over and talk to the doc!'"[96]

limited funding for such work in the extramural program to the cell lines then in existence.[21]

Not every feature of Bush's policy was new. The Clinton administration and the Congress had ordered that the government not fund the creation of human embryos solely for research purposes, although, at the time, the reason for the restriction was to prohibit human cloning.[46,97]* However, funding was allowed for research performed on cell lines created elsewhere and not funded by the federal government. Harold Varmus, when he was director of the NIH from 1993 to 1999, said, "we should be supporting research on the derivation of stem cells from spare embryos. It may not be politically appropriate to do that at this point. I think any ethical evaluation has to take into account the consequences of not doing research that would benefit living people who have serious diseases now or in the future."[98]

The policy from 2001 was that federally funded stem cell research had to be limited to studies on cell lines created before August 9, 2001, and prohibited studies on any line derived afterwards. Since then, many embryonic cell lines have been created, unsupported by federal funding, that could be used, and are being used, for human research. Research has recently suggested that stem cells can be developed from human tissue other than embryos, which, if possible, could obviate the rationale for Bush's policy.[99]

The Bush decision was not well received by many at the NIH. Gerald Fischbach found the rules "so repressive. I knew that my days at the NIH were numbered."[66] Fischbach resigned as director of the National Institute of Neurological Disorders and Stroke later that year.

Varmus, in responding to the efforts of state governments to fund federally prohibited research on new cell lines, said:[46]

> The emergence of several state initiatives to use public funds for research ... reveals some of the advantages of having national review and funding mechanisms for research. NIH provides a

---

* Animals have been cloned, the most famous being the sheep named Dolly. The technique involves removing the nucleus, which contains the animal's genetic material, from the egg of the female recipient. The genetic material from the animal to be cloned is inserted into the recipient's egg. The cloned cell is then transplanted into the recipient animal's uterus. The animal that is born—when the technique works as it does in only a minority of attempts—has the identical genetic characteristics of the donor.

fairly level playing field, but we are moving towards a heterogeneous collection of state-based research opportunities, with scientists in some states running the risk of criminal charges for doing what other states would pay them handsomely to do.

NIH Stem Cell Task Force

One year after the Bush directive, NIH director Elias Zerhouni asked James Battey, director of the National Institute on Deafness and Other Communication Disorders, to chair a task force on embryonic stem cell research.[100] "Since my institute was not funding much work using human embryonic stem cells, Elias saw me as a 'neutral factor,'" Battey said. "Since then, however, I've had to become much better informed about the subject."[25]

Accordingly, Battey became an important source of information about stem cell issues for the press and media. Since each of his replies must be approved by officials at the department of Health and Human Services and, in some cases, he presumes, by the White House, Battey has instructed his secretary to "always respond that I'll call back whether or not I'm really available. The request goes to our office of communications and public liaison and often to DHHS. Absent a 'go' signal, I don't go. Sometimes they can't wait and publish the story anyway."[25] Others fault the Department of Health and Human Services for responding slowly to requests from the NIH for clearance.[8]

Speaking as a scientist and concerned citizen, Battey said, "I have never defended the logic of the stem cell controversy. If unused embryos have been discarded as medical waste for years, why can't they be dedicated to scientific and medical purposes? The politicians are interfering with scientific progress for political purposes."[97]

The Director Responds

On March 19, 2007, Elias Zerhouni appeared before the Senate Health Appropriations Subcommittee and requested that the Bush policy be reversed. Zerhouni said:[101]

> From my standpoint, it is clear today that American science will be better served, and the nation will be better served, if we let our scientists have access to more stem cell lines. We cannot, I would think, be second-best in this area. I think it is important for us not

Chapter Eight: Controversies

to fight with one hand tied behind our back here, and NIH is key to that.

In 2004, Zerhouni had begun to express reservations about the president's policy and its impact on research. In a public letter to Congress, he stated that from the purely scientific standpoint, more stem cell lines eligible for federal funding would be useful, but he refrained from criticizing directly the policy recognizing the right of the president and Congress to make policy. On March 19, 2007, Elias Zerhouni went further when he appeared before the Senate Health Appropriations Subcommittee and requested that the Bush policy be reversed.[101]

Zerhouni explained that the effectiveness of the other methods of harvesting stem cells is "overstated. We do not know at this point where the breakthrough will come from. All angles in stem cell research should be pursued." Although several states are raising money for stem cell research, Zerhouni, like Varmus, said that the "NIH was better suited to conduct such long-term and difficult projects. There is no state that can really provide the depth and oversight and stimulation of this research over the long run."[101]

Prominent investigators welcomed Zerhouni's remarks. One suggested that the director may have taken a political risk by articulating a position so clearly at odds with that of the White House. "He's very careful about what he said, and fairly conservative in his policy statements," said Fred Gage of the Salk Institute in La Jolla, California. "My guess is this was a very well-thought-out statement. If it does put him at risk, all the more reason to respect his judgment."[101] Senator Tom Harkin, chairman of the subcommittee before which Zerhouni was appearing, called his remarks "very profound and courageous."[101]*

---

* Zerhouni writes, "I took the risk because I was very concerned that the recent development of induced pluripotent stem cells ... [was] being publicly and privately used by opponents of stem cell research inside and outside the Administration and Congress as a reason to no longer fund embryonic stem cell research, a move I wanted to preempt. I had already indicated to Secretary Leavitt and the President my strong opposition to the representations made in that regard by White House and HHS staffers. Knowing the ways of Washington, I decided that the best way to fend this off was to go public to prevent such an outcome. This did cost me the ire of some in the Administration but the President had always appreciated my candor in giving him my professional opinions on many topics even in disagreement and he let it be known that he continued to support me despite my public statements. I was both pleased and surprised that I was allowed to finish my term as NIH director after such a public disagreement with the President."[21]

The Obama Policy

On March 10, 2009, President Obama lifted the Bush administration's strict limits on human embryonic stem research as part of his pledge to "make scientific decisions based on facts, not ideology."[102]

CHAPTER NINE

Conclusions

No matter the administrative and budgetary problems associated with any governmental agency, the NIH survives and prospers because the public and the Congress recognize the valuable work its scientists perform. As I have tried to describe, however, the NIH has its problems, and being a part of the federal government clearly prevents instituting what one might see as obvious solutions.

Administrative Structure

The NIH has prospered with its separate institutes and centers, although in many cases, one has difficulty understanding why some of the institutes contain the medical specialties they do. The National Cancer Institute, the first of the institutes and the one with the largest budget, deals with only one "illness," although the NCI lists 230 different types of cancer.

Each of three institutes—National Institute of Diabetes and Digestive and Kidney Diseases, National Institute of Arthritis, Musculoskeletal, and Skin Diseases, and National Heart, Lung, and Blood Institute—have responsibility to support three major disease categories. Within each are many different illnesses plus basic research that in some way can be said to be related to an illness that falls within the purview of that institute.

Often there is some logic in how the multi-disease institutes were structured. NHLBI is a good example. It started as the National Heart Institute, later the "Lung" was added, and finally "Blood." Since the heart pumps blood to the lungs, one can see some logic in this arrangement. The connection between the organs and diseases assigned to National Institute of Diabetes and Digestive and Kidney Diseases is not so obvious, although one could rationalize the arrangement since diabetes contributes to the development of both gastrointestinal and kidney diseases.

One can understand why Harold Varmus, some of his predecessors, and the organizations that have reviewed the NIH advocate reorganizing this alphabet soup. These efforts have made little headway. That the NIH is a governmental and not a private agency accounts for much of this. We must remember how the pressure exerted by advocates influences much of the structure of the NIH, which in some cases supersedes what logic would dictate.

Though saving money is a prime objective of consolidating institutes, no one knows how much would be saved.[1] At least, according to a provision of the recent Reauthorization Act, no more institutes will be formed.

Intramural versus Extramural Research Programs

Biomedical science has become progressively more dependent on the NIH to fund much of the best research being conducted in the nation's medical centers and institutes. Though some in the extramural community indulge in the fantasy that the money spent on intramural programs should be assigned to them, one has only to catalogue the discoveries since World War II made by the scientists on the Bethesda and affiliated campuses to realize how valuable the intramural program is.

Careers in the intramural program offer advantages not available in medical schools and research institutes. Chief among these is that intramural investigators do not have to write grants to support the expenses of the laboratories and salaries for the investigator and associates. Their work is reviewed, however, on a regular basis, and a poor review can lead to reduction of support or even closure of the laboratory. These reviews, however, are retrospective in that the investigator's past work is studied, whereas the grant mechanism in the extramural program is prospective in that the applicant is evaluated on the merits of the work to be performed in the future.

Although many have said that working in the intramural program not so popular a career as it once was, data to establish this assumption are hard to find. It is true that there are many more institutions that offer excellent opportunities to work in biomedical research than when the NIH was young, in the decades following World War II.

Finances

Beginning in fiscal year 2004 when the fiscal doubling ended, the government's support of the NIH, when factoring in the effects of inflation, decreased in value each year. The medical schools, teaching hospitals, and research institutes whose investigators receive most of the money through the extramural research program complained vigorously. During the doubling, many of these institutions had added faculty, trainees, and physical facilities to be partially funded by the NIH where annual increases in the NIH budget would help pay for these improvements. Instead, Congress and the administration gave the NIH less purchasing power each year, and the usual recipients of the extramural funds had to scramble to balance their budgets.

These events showed, as if most of us didn't know, how dependent academic medicine had become on the NIH. Many blamed the bad times on Republican control of the Congress and, more specifically the White House, where, it seemed, there was significantly less enthusiasm than in former administrations for supporting scientific research in general and biomedical research in particular. Under the previous administration, the Congress, even though controlled by the Republicans for a majority of the years that Bill Clinton was president, supported the NIH to an unparalleled extent, culminating in the doubling. Different priorities, including wars in Afghanistan and Iraq at a time when taxes were being significantly reduced, supported what many believe was the Bush administration's bias against investing more money in science and biomedicine.

As I write this chapter in early 2009, America is in the midst of one of the most troubling economic periods in decades. Despite this, a brighter day for the support of biomedical science has arrived with the administration of Barack Obama. In February, as part of the administration-sponsored American Recovery and Reinvestment Act, or "stimulus fund," the NIH budget was increased, thanks in part to the successful political maneuvering of Senator Arlen Specter, a longtime and energetic supporter of biomedical research.²*

If the patterns of the past can be projected into the future, doing bioscience will cost more each year, thanks to the development of new and often expensive equipment and supplies needed to solve medical problems. Nevertheless, one can be certain, given the presence in our country of so many talented scientists, that biomedical science will continue to answer fundamental questions about the nature of disease, the knowledge of which can be translated into improved clinical care.

---

* The money, much of which the NIH will try to spend by September, 2009, will be available through the end of fiscal year 2010. See http://www.nih.gov/about/director/02252009statement_arra.htm for details about how the NIH plans to spend the funds.

APPENDIX A: INTERVIEWEES

Past and present titles at the National Institutes of Health are listed first. NIH institutes and centers are given their current names. Titles without dates are those the interviewees held when interviews were conducted, followed by current positions, if different or relevant. Locations are in Bethesda, Maryland, unless otherwise indicated.

Abbreviations:
CC = Clinical Center
CEO = Chief Executive Officer
CFO = Chief Financial Officer
COO = Chief Operating Officer
DHHS = Department of Health and Human Services
EVP = Executive Vice-President
HMS = Harvard Medical School
IOM = Institute of Medicine
NIH = National Institutes of Health
UMSOM = University of Maryland School of Medicine
VP = Vice President

Alexander, Duane F., M.D.	Director, National Institute of Child Health and Human Development (NICHD), NIH
Alter, Harvey J., M.D.	Chief, Infectious Diseases Section, Department of Transfusion Medicine, Clinical Center (CC), NIH
Alving, Barbara M. N., M.D.	Director, National Center for Research Resources (NCRR), NIH

(continued)

(Continued)

Angell, Marcia, M.D.	Senior Lecturer on Social Medicine, HMS, Boston, MA; Interim Editor-in-Chief, *New England Journal of Medicine* (1999–2000), Boston, MA
Arai, Andrew E., M.D.	Senior Investigator, Laboratory of Cardiac Energetics, National Heart, Lung, and Blood Institute (NHLBI), NIH
Ayres, Elaine J.	Assistant Director for Ethics and Technology Development, CC, NIH
Baltimore, David, Ph.D.	President Emeritus, Robert Andrews Millikan Professor of Biology, California Institute of Technology, Pasadena, CA
Barros, Colleen F.	CFO, Deputy Director for Management, NIH
Bartrum, John J.	Associate Director for Budget, NIH
Battey, Jr., James F., M.D., Ph.D.	Director, National Institute on Deafness and Other Communication Disorders (NIDCD), NIH
Baum, Carolyn A.	Program Analyst, Office of Federal Advisory Committee Policy, NIH
Baum, Stanley, M.D.	Editor-in-Chief, *Academic Radiology*; President, Academy of Radiology Research (1997–1999); Chairman, Department of Radiology, University of Pennsylvania School of Medicine, Philadelphia, PA (1975–1996)
Baumgardner, William A., M.D.	Cardiac Surgeon-in-Charge, Johns Hopkins Hospital; Vincent L. Gott Professor of Surgery, Johns Hopkins University School of Medicine
Benz, Jr., Edward J., M.D.	Chair, Advisory Board for Clinical Research, CC, NIH; President, Dana-Farber Cancer Institute, Boston, MA
Berg, Jeremy M., Ph.D.	Director, National Institute of General Medical Sciences(NIGMS), NIH
Berman, Brian M., M.D.	Director, Center for Integrative Medicine, UMSOM, Baltimore, MD
Boyce, Thomas M.	Program Manager, Electronic Research Administration (eRA), NIH (2005–2008); Director, Information Services and Deputy Chief Information Officer, U.S. Nuclear Regulatory Commission

Braunwald, Eugene, M.D.	Chief, Section of Cardiology, Clinic of Surgery; Chief, Cardiology Branch; Clinical Director, National Heart, Lung and Blood Institute, (NIHLB), NIH (successive appointments from 1958–1968); Distinguished Hersey Professor of Medicine, HMS; Chairman, TIMI Study Group and Senior Physician, Brigham and Women's Hospital, Boston, MA
Bridbord, Kenneth, M.D.	Director, Division of International Training and Research, Fogarty International Center (FIC), NIH
Brieger, Gert H., M.D., Ph.D.	Director, Department of History of Science, Medicine and Technology, Johns Hopkins University and Johns Hopkins University School of Medicine (1984–2002)
Briggs, Josephine P., M.D.	Director, National Center for Complementary and Alternative Medicine (NCCAM), NIH
Broder, Samuel, M.D.	Director, National Cancer Institute (NCI), NIH (1988–1995); Chief Medical Officer, Celera Genomics Corporation, Rockville, MD
Brown, David M., M.D.	Staff Writer, *The Washington Post*, Washington, DC
Brown, Michael S., M.D.	Clinical Associate, National Institute of Diabetes and Digestive and Kidney Diseases (NIDDK) (1968–1970); post doctoral fellow, National Heart, Lung, and Blood Institute (NHLBI) (1970–1971), NIH; Director, Jonsson Center for Molecular Genetics, University of Texas Southwestern Medical Center, Dallas, TX; Nobel Prize in Physiology or Medicine, 1985
Brugge, Joan Siefert, Ph.D.	Chair, Department of Cell Biology, HMS, Boston, MA
Burklow, John T.	Associate Director for Communications and Public Liaison, NIH
Butler, Robert N., M.D.	Director, National Institute on Aging (NIA), NIH (1976–82); President and CEO, International Longevity Center–USA; Professor of Geriatrics, Henry L. Schwartz Department of Geriatrics and Adult Development, Mount Sinai Medical Center, New York, NY

(*continued*)

(Continued)

Byars, Sara R.	Senior Communications Advisor, CC, NIH
Califf, Robert M., M.D.	Director, Duke Clinical Research Institute and Vice-Chancellor for Clinical Research, Duke University Medical Center, Durham, NC
Caplan, Arthur L., Ph.D.	Chair, Department of Medical Ethics, and Director, Center for Bioethics, University of Pennsylvania School of Medicine, Philadelphia, PA
Casey, Kevin	Senior Director of Federal and State Relations, Harvard University, Cambridge, MA
Cassell, Gail H., Ph.D.	VP, Scientific Affairs, and Distinguished Lilly Research Scholar for Infectious Diseases, Eli Lilly and Company, Indianapolis, IN
Charney, Dennis S., M.D.	Chief, Mood and Anxiety Disorders Research Program and Experimental Therapeutics and Pathophysiology Branch, National Institute of Mental Health (NIMH), NIH (2000–2004); Dean for Academic and Scientific Affairs, Mount Sinai School of Medicine; Senior Vice President for Health Sciences, Mount Sinai Medical Center, New York, NY
Chen, Philip S., Ph.D.	Various positions, including several in the Office of the Director (1956–1959, 1967–2006), NIH
Cohen, Paula	Public Affairs Specialist, Office of Communications and Public Liaison, NIH
Coller, Barry S., M.D.	Vice Chair, Advisory Board for Clinical Research, NIH; VP for Medical Affairs, Physician-in-Chief, Rockefeller University, New York, NY
Collins, Francis S., M.D., Ph.D.	Director, National Human Genome Research Institute (NHGRI), (1993–2008), Director (2009-) NIH
Collins, Michael T., M.D.	Chief, Skeletal Clinical Studies Unit, National Institute of Dental and Craniofacial Research (NIDCR), NIH
Corn, Milton, M.D.	Director, Division of Extramural Programs, and Associate Director, National Library of Medicine (NLM), NIH
Croyle, Robert T., Ph.D.	Director, Division of Cancer Control and Population Sciences, National Cancer Institute (NCI), NIH

Culliton, Barbara J.	Journalist, Editor-in-Chief, *Science* Magazine
Demsey, Anthony, Ph.D.	Associate Director for Research Administration, National Institute of Biomedical Imaging and Bioengineering (NIBIB), NIH
DeVita, Vincent T., Jr., M.D.	Director, National Cancer Institute (NCI) (1980–1988), NIH; Chairman, Yale Cancer Center Advisory Board, Yale University, New Haven, CT
Dickler, Robert M.	Chief Health Care Officer, Association of American Medical Colleges (AAMC), Washington, DC
Ehrenfeld, Elvera (Ellie), Ph.D.	Head, Picornavirus Replication Section, Laboratory of Infectious Diseases, National Institute of Allergy and Infectious Diseases (NIAID); Director, Center for Scientific Review (CSR) (1997–2003), NIH
Emanuel, Ezekiel J., M.D., Ph.D.	Chairman, Department of Clinical Bioethics, CC, NIH
English, Joseph T., M.D.	Clinical Associate, National Institute of Mental Health, (NIMH), NIH (1961–1962); Professor, Chairman and Associate Dean, Department of Psychiatry and Behavioral Sciences, New York Medical College, Valhalla, NY; System Chairman, Department of Psychiatry and Behavioral Sciences, St. Vincent Catholic Medical Centers, New York, NY
Fauci, Anthony S., M.D.	Director, National Institute of Allergy and Infectious Diseases (NIAID), NIH
Fee, Elizabeth, Ph.D.	Chief, History of Medicine Division, National Library of Medicine (NLM), NIH
Finkel, Toren, M.D., Ph.D.	Chief, Translational Medicine Branch, National Heart, Lung, and Blood Institute (NHLBI), NIH
Fischbach, Gerald D., M.D.	Director, National Institute of Neurological Disorders and Stroke (NINDS), NIH (1998–2001), John E. Borne Professor of Medical and Surgical Research, Columbia University College of Physicians and Surgeons, New York, NY; Scientific Director, The Autism Initiative, Simons Foundation, New York, NY

(continued)

(Continued)

Fisher, Suzanne E., Ph.D.	Director, Division of Receipt and Referral, Center for Scientific Review (CSR), NIH
FitzGerald, Garret A., M.D.	Robert L. McNeil, Jr., Professor in Translational Medicine and Therapeutics; Chairman, Department of Pharmacology, University of Pennsylvania School of Medicine, Philadelphia, PA
Fleisher, Thomas A., M.D.	Chief, Department of Laboratory Medicine, CC, NIH
Folkers, Richard A.	Director, Office of Media Relations, National Cancer Institute (NCI), NIH
Ford, Daniel E., M.D.	Vice Dean for Clinical Investigation, Johns Hopkins University School of Medicine, Baltimore, MD
Fox, E. Brooke	Archivist, Documents and Photographs Collections, NIH
Fradkin, Judith E., M.D.	Director, Division of Diabetes, Endocrinology, and Metabolic Diseases, National Institute of Diabetes and Digestive and Kidney Diseases (NIDDK), NIH
Fraser-Liggett, Claire M., Ph.D.	Various positions, National Institute of Neurological Disorders and Stroke (NINDS) and National Institute on Alcohol Abuse and Alcoholism (NIAAA), NIH (1985–1992); Director, Institute for Genome Sciences, UMSOM, Baltimore, MD
Gallin, John I., M.D.	Director, CC, NIH
Gallo, Robert C., M.D.	Director, Institute of Human Virology, UMSOM, Baltimore, MD
Gerber, Lynn H., M.D.	Chief, Department of Rehabilitation Medicine, CC, NIH (1976–2005); Director, Center for Chronic Illness and Disability, and Professor of Rehabilitation Science, George Mason University, Fairfax, VA
Germain, Ronald N., M.D., Ph.D.	Deputy Chief, Laboratory of Immunology, National Institute of Allergy and Infectious Diseases (NIAID), NIH
Gibbons, Don L.	Associate Dean for Public Affairs, HMS, Boston, MA (1996–2008); Chief Communications Officer, California Institute for Regenerative Medicine, San Francisco, CA, beginning 2008

Glass, Roger I., M.D., Ph.D.	Director, Fogarty International Center (FIC), NIH
Goldman, Howard H., M.D., Ph.D.	Research Psychiatrist, Division of Biometry and Epidemiology (1978–1980) and Assistant Institute Director, National Institute of Mental Health (NIMH) (1983–1985), NIH; Professor of Psychiatry, UMSOM, Baltimore, MD
Goldman, Lee, M.D.	Executive VP for Health and Biomedical Sciences, and Dean of the Faculties of Health Sciences and of Medicine, Columbia University College of Physicians and Surgeons, New York, NY
Goldstein, Joseph L., M.D.	Clinical Associate, National Heart, Lung, and Blood Institute, NIH (1966–1968); Chairman, Department of Molecular Genetics, University of Texas Southwestern Medical Center, Dallas, TX; Nobel Prize in Physiology or Medicine, 1985
Gorden, Phillip, M.D.	Senior Investigator, former Director, National Institute of Diabetes and Digestive and Kidney Diseases (NIDDK), NIH; Clinical Professor of Medicine, Uniformed Services Medical School
Gormley, Maureen E., R.N.	Chief Operating Officer, CC, NIH
Gottesman, Michael M., M.D.	Deputy Director for Intramural Research, NIH
Grady, Patricia A., B.S.N., Ph.D.	Director, National Institute of Nursing Research (NINR), NIH
Graeff, Alan S.	Deputy Director, National Center for Biotechnology Information (NCBI), National Library of Medicine (NLM), NIH; Director, Center for Information Technology (CIT), NIH (1998–2005)
Gragnolati, Brian A.	President and CEO, Suburban Hospital Healthcare System
Greenberg, Daniel S.	Freelance Writer, Chevy Chase, MD
Greenberger, Phyllis E., M.S.W.	President and CEO, Society for Women's Health Research, Washington, DC
Greenwald, Harriet R.	Executive Director, NIH Alumni Association and Editor, NIHAA *Update*, NIH (1988–2007)

(*continued*)

(Continued)

Grob, Gerald N., Ph.D.	Professor of History Emeritus, Rutgers, The State University of New Jersey, New Brunswick, NJ
Groft, Stephen C., Pharm. D.	Director, Office of Rare Diseases, Office of the Director (OD), NIH
Hall, Zach W., Ph.D.	Director, National Institute of Neurological Disorders and Stroke (NINDS), NIH (1994–1997)
Harden, Victoria A., Ph.D.	Special Volunteer, Office of NIH History; former Director, DeWitt Stetten, Jr., Museum of Medical Research, NIH; Scholar-in-Residence, American University, Washington, DC
Harkins, Barbara	Archivist/Librarian, Office of NIH History, NIH
Haseltine, Florence P., M.D.	Director, Center for Population Research, National Institute of Child Health and Human Development (NICHD), NIH
Hastings, Clare, R.N., Ph.D.	Chief, Nursing and Patient Care Services, CC, NIH
Hayward, Anthony R., M.D., Ph.D.	Director, Division for Clinical Research Resources, National Center for Research Resources (NCRR), NIH
Healy, Bernadine P., M.D.	Director, NIH (1991–1993); Columnist and Health Editor, U.S. News and World Report.
Henderson, David K., M.D.	Deputy Director for Clinical Care, CC, NIH
Hodes, Richard J., M.D.	Director, National Institute on Aging (NIA), NIH
Hollingsworth, J. Rogers, Ph.D.	Professor of Sociology and History Emeritus, University of Wisconsin, Madison, WI
Horvath, Keith A., M.D.	Director, Cardiothoracic Surgery Research Program, National Heart, Lung, and Blood Institute (NHLBI), NIH; Chief, Cardiothoracic Surgery, Suburban Hospital
Hyman, Steven E., M.D.	Director, National Institute of Mental Health (NIMH), NIH (1996–2001); Provost, Harvard University, Cambridge, MA
Iglehart, John K.	National Correspondent, *New England Journal of Medicine*, Boston MA
Insel, Thomas R., M.D.	Director, National Institute of Mental Health (NIMH), NIH

Appendix A: Interviewees

Jacobs, Joseph J., M.D.	Director, Office of Alternative Medicine (1992–1994), NIH; Associate Medical Director, Abbott Molecular, Des Plaines, IL
Jaffe, Holli Beckerman, J.D.	Director, Ethics Office, NIH
James, Stephen P., M.D.	Director, Division of Digestive Diseases and Nutrition, National Institute of Diabetes and Digestive and Kidney Diseases (NIDDK), NIH
Jones, John F., Ph.D.	Acting Director, Center for Information Technology (CIT), NIH
Katz, Stephen I., M.D., Ph.D.	Director, National Institute of Arthritis and Musculoskeletal and Skin Diseases (NIAMS), NIH
Kelley, Rebecca L.	Director, Office of Extramural Research (OER), NIH
Kelley, William N., M.D.	Clinical Associate, National Institute of Arthritis and Metabolic Diseases, NIH (1965–1967); Professor of Medicine, University of Pennsylvania School of Medicine, Philadelphia, PA
Kent, Kenneth M., M.D., Ph.D.	Chief of Cardiology, Suburban Hospital
Killen, John Y., M.D.	Director, Office of International Health Research (OIHR); Acting Deputy Director and Acting Director, Division of Extramural Research and Training, National Center for Complementary and Alternative Medicine (NCAAM), NIH
Kirby, Kevin E., Ed.D.	Executive Officer, National Institute of Neurological Disorders and Stroke (NINDS), NIH (1998–2002); VP for Administration, Rice University, Houston, TX
Kirschner, Marc W., Ph.D.	Chair, Department of Systems Biology, HMS, Boston, MA
Kirschstein, Ruth L., M.D.	Senior Advisor to the Director; Acting Director, National Center on Complementary and Alternative Medicine (2006–2008), NIH; Acting Director, NIH (1993, 2001–2002); Director, National Institute of General Medical Sciences (NIGMS) (1973–1993), NIH

(*continued*)

(Continued)

Klag, Michael J., M.D.	Dean, Johns Hopkins Bloomberg School of Public Health, Baltimore, MD
Klausner, Richard D., M.D.	Director, National Cancer Institute (NCI), NIH (1995–2001)
Klein, Harvey G., M.D.	Director, Department of Transfusion Medicine, CC, NIH
Kleinman, Dushanka V., D.D.S.	Associate Dean for Research and Academic Affairs and Professor of Epidemiology and Biostatistics, College of Health and Human Performance, University of Maryland, College Park, MD
Kolberg, Rebecca	Deputy Chief, Communications and Public Liaison Branch, National Human Genome Research Institute (NHGRI), NIH
Korn, David, M.D.	Research Associate, National Institute of Arthritis and Metabolic Disorders (NIAMD), NIH (1961–1963); Senior VP for Biomedical and Health Sciences Research, Association of American Medical Colleges, Washington, DC (1997–2008); Vice Provost for Research, Harvard University, Cambridge, MA
Krakower, Jack, Ph.D.	Senior Director, Medical School Financial and Administrative Affairs, Association of American Medical Colleges (AAMC), Washington, DC
Lander, Eric S., D.Phil.	Director, Eli and Edythe L. Broad Institute of MIT and Harvard, Cambridge, MA
Landis, Story C., Ph.D.	Director, National Institute of Neurological Disorders and Stroke (NINDS), NIH
Landsman, David	Senior Investigator and Branch Chief, Computational Biology Branch, National Library of Medicine (NLM), NIH
Lauer, Michael S., M.D.	Director, Division of Prevention and Population Sciences, National Heart, Lung, and Blood Institute (NHLBI), NIH
Law, Catherine	Science Writer/Press Team Leader, National Center for Complementary and Alternative Medicine (NCCAM), NIH
Lederhendler, Israel I., Ph.D.	Director, Division of Information Services, Office of Extramural Research, NIH

Lee, Marvin	Chief, Special Programs Branch, Office of Human Resources, NIH
Lee, Philip R., M.D.	Assistant Secretary for Health, Department of Health and Human Services (1993–1997); Consulting Professor, Program in Human Biology, Stanford University, Stanford CA
Lenardo, Michael J., M.D.	Chief, Molecular Development of the Immune System Section, National Institute of Allergy and Infectious Diseases (NIAID), NIH
Lenfant, Claude, M.D.	Director, National Heart, Lung, and Blood Institute (NHLBI), NIH (1982–2003)
Leshner, Alan I., Ph.D.	Director, National Institute on Drug Abuse (NIDA) (1994–2001); President, National Association for the Advancement of Science, Washington, DC
Levine, Arthur S., M.D.	Various positions, National Cancer Institute (NCI) (1966–1982), NIH; Scientific Director, National Institute of Child Health and Human Development (NICHD) (1982–1998), NIH; Senior Vice Chancellor for Health Sciences, and Dean, University of Pittsburgh School of Medicine, Pittsburgh, PA
Levine, Myron M., M.D.	Director, Center for Vaccine Development, University of Maryland School of Medicine, Baltimore, MD.
Li, Ting-Kai, M.D.	Director, National Institute on Alcohol Abuse and Alcoholism (NIAAA), Rockville, MD
Lindberg, Donald A.B., M.D.	Director, National Library of Medicine (NLM), NIH
Liotta, Lance A., M.D., Ph.D.	Chief, Laboratory of Pathology, National Cancer Institute (NCI), NIH (1982–2005); University Professor and Professor of Life Sciences; Co-Director, Center for Applied Proteomics and Molecular Medicine, George Mason University, Manassas, VA
Lipman, David J., M.D.	Director, National Center for Biotechnology Information, National Library of Medicine (NLM), NIH
Luckett, Donald N.	Communications Director, Center for Scientific Review (CSR), NIH

(*continued*)

(Continued)

Mahmoud, Adel, M.D.	Senior Policy Analyst, Woodrow Wilson School and Molecular Biology, Lecturer with the rank of Professor in Molecular Biology, Princeton University, Princeton, NJ
Major, Christine M.	Director, Office of Human Resources and Office of Strategic Management Planning, NIH
Mandel, H. George, Ph.D.	Professor of Pharmacology and Physiology, George Washington University School of Medicine, Washington, DC
Marks, Andrew R., M.D.	Chairman, Department of Physiology and Cellular Biophysics, Columbia University College of Physicians and Surgeons, New York, NY
Martensen, Robert L., M.D., Ph.D.	Director, Office of NIH History and Museum, NIH
Martin, Joseph B., M.D., Ph.D.	Dean of the Faculty of Medicine, HMS (1997–2007), Boston, MA
Masur, Henry, M.D.	Chief, Department of Critical Care Medicine, CC, NIH
Mazzaschi, Anthony J.	Senior Director, Scientific Affairs, Association of American Medical Colleges (AAMC), Washington, DC
McIntosh, Charles L., M.D., Ph.D.	Chief Medical Science and Technology Officer, Cook Group Incorporated, Bloomington, IN
McKhann, Guy M., M.D.	Research Associate (1957–1960), Associate Director for Clinical Research, Acting Clinical Director, Advisor to the Director, National Institute of Neurological Disorders and Stroke (NINDS), NIH (successive appointments from 2000 to present); Professor of Neurology and Neuroscience, Johns Hopkins University, Baltimore, MD
Menikoff, Jerry, M.D., J.D.	Director, Office of Human Subjects Research (OHSR), NIH
Milgram, Sharon L., M.D.	Director, Office of Intramural Training and Education, NIH
Miller, Edward D., M.D.	Dean, Johns Hopkins University School of Medicine; CEO, Johns Hopkins Medicine, Baltimore, MD

Appendix A: Interviewees

Mitchell, Robert F.	Director, Research Services, Department of Medicine, UMSOM, Baltimore, MD
Myers, Louise M.	Chief, Budget Formulation, Presentation, and Execution Branch, Office of Budget, NIH
Nabel, Elizabeth G., M.D.	Director, National Heart, Lung, and Blood Institute (NHLBI), NIH
Nabel, Gary J., M.D., Ph.D.	Director, Vaccine Research Center, National Institute of Allergy and Infectious Diseases (NIAID), NIH
Nathan, David G., M.D.	Clinical Associate, National Cancer Institute, NIH (1956–1958); President Emeritus, Dana-Farber Cancer Institute, Robert A. Stranahan Distinguished Professor of Pediatrics and Professor of Medicine, HMS, Boston, MA
Niederhuber, John E., M.D., Ph.D.	Director, National Cancer Institute (NCI), NIH
Noguchi, Constance Tom, Ph.D.	Dean, Foundation for Advanced Education in the Sciences, Inc.(FAES) Graduate School, NIH
Notkins, Abner L., M.D.	Chief, Experimental Medicine Section, National Institute of Dental and Craniofacial Research (NIDCR); Scientific Director, Intramural Research Program, National Institute of Dental and Craniofacial Research (NIDCR)(1984–1992), NIH
Ognibene, Frederick P., M.D.	Director, Office of Clinical Research Training and Medical Education, CC, NIH
Oldfield, Edward H., M.D.	Neurosurgeon, National Institute of Neurological Disorders and Stroke (1981–2007); Professor of Neurosurgery and Internal Medicine, University of Virginia, Charlottesville, VA
Pardes, Herbert, M.D.	Director, National Institute of Mental Health, (NIMH), NIH (1978–1984); President and CEO, New York-Presbyterian Hospital, New York, NY
Passamani, Eugene R., M.D.	Senior VP for Research and Education, Suburban Hospital Healthcare System
Patterson, Amy P., M.D.	Director, Office of Biotechnology Activities, Office of the Director, NIH

(*continued*)

(Continued)

Paul, William E., M.D.	Chief, Laboratory of Immunology, National Institute of Allergy and Infectious Diseases, (NIAID) and Senior Investigator, NIH
Pelis, Kimberly A., Ph.D.	Speechwriter for Elias Zerhouni, NIH
Pettigrew, Roderic I., M.D., Ph.D.	Director, National Institute of Biomedical Imaging and Bioengineering (NIBIB), NIH
Pinn, Vivian W., M.D.	Director, Office of Research on Women's Health, NIH
Pizzo, Philip A., M.D.	Clinical Associate, Senior Investigator, Head, Infectious Diseases Section; Chief of Pediatrics successive appointments in the National Cancer Institute (NCI), NIH (1974–1996); Dean, Stanford University School of Medicine, Stanford, CA
Pope, Andrew M.	Director, Board on Health Sciences Policy, IOM
Porter, Amy McGuire	Executive Director, Foundation for the National Institutes of Health, Bethesda, MD
Porter, John E.	Partner, Hogan and Hartson; Member of Congress (R-IL) (1980–2001)
Potts, Jr., John T., M.D.	Trainee and Investigator, National Heart Institute, NIH (1959–1968); Jackson Distinguished Professor of Clinical Medicine, HMS, Boston, MA
Ramm, Louise E., Ph.D.	Deputy Director, National Center for Research Resources (NCRR), NIH
Read, Elizabeth J., M.D.	Medical Director and Chief, Cell Processing Section, Department of Transfusion Medicine, CC, NIH (1995–2006); Director, Cell and Tissue Therapies, Blood Systems Research Institute, San Francisco, CA
Relman, Arnold S., M.D.	Professor of Medicine Emeritus, HMS; Editor-in-Chief Emeritus, *New England Journal of Medicine*, Cambridge, MA
Rettig, Richard A., Ph.D.	Adjunct Senior Social Scientist, RAND Corporation, St. Augustine, FL (semi-retired)

Reynolds, Craig W., Ph.D.	Director, Office of Scientific Operations, and Associate Director, National Cancer Institute (NCI), NIH
Rodgers, Griffin P., M.D.	Director, National Institute of Diabetes and Digestive and Kidney Diseases (NIDDK)
Rogers, Paul G.	Partner, Hogan and Hartson, Washington; Member of Congress (D-Fla.) (1955–1979).
Rogers, Terry B., Ph.D.	Professor of Biochemistry and Molecular Biology, Director, M.D./Ph.D. Program, UMSOM, Baltimore, MD
Rohrbaugh, Mark L., Ph.D., J.D.	Director, Office of Technology Transfer, NIH
Rosenberg, Steven A., M.D., Ph.D.	Chief of Surgery, National Cancer Institute (NCI), NIH
Rosenstein, Donald L., M.D.	Clinical Director and Chief, Psychiatry Consultation-Liaison Service, National Institute of Mental Health (NIMH), NIH
Roth, Carl A., Ph.D.	Associate Director for Scientific Program Operation, National Heart, Lung, and Blood Institute (NHLBI), NIH
Rowland, Lewis P., M.D.	Clinical Associate, National Institute of Neurological Disorders and Stroke (NINDS), NIH (1953–1954); Professor of Neurology, Columbia University College of Physicians and Surgeons, New York, NY
Rubenstein, Arthur H., M.B.B.Ch.	Dean, University of Pennsylvania School of Medicine; EVP for the Health System, University of Pennsylvania, Philadelphia, PA
Ruiz Bravo, Norka, Ph.D.	Deputy Director for Extramural Research, NIH
Sampat, Blaven N., Ph.D.	Assistant Professor of Health Policy and Management, Mailman School of Public Health, Columbia University, New York, NY
Sanders, Charles A., M.D.	Chairman, Board of Directors, Foundation for the National Institutes of Health, Bethesda, MD
Scarpa, Antonio, M.D., Ph.D.	Director, Center for Scientific Review (CSR), NIH
Schaffer, Walter T. Ph.D.	Senior Scientific Advisor for Extramural Research, NIH

(*continued*)

(Continued)

Schechter, Alan N., M.D.	Chief, Molecular Medicine Branch, National Institute of Diabetes and Digestive and Kidney Diseases (NIDDK), NIH
Schimpff, Stephen C., M.D.	Member (1996–2003) and Chair (2000–2003), Board of Governors, CC, NIH; Professor of Medicine and Public Policy, University of Maryland, Baltimore and College Park, MD
Schwartz, David A., M.D.	Director, National Institute of Environmental Health Sciences (NIEHS), NIH (2005–2007); Director, Pulmonary Division and Genetics Center, National Jewish Medical and Research Center, Denver, CO
Shamoo, Adil E., Ph.D.	Professor of Biochemistry and Molecular Biology, UMSOM, Baltimore, MD
Shapiro, Bert I., Ph.D.	Program Director, Medical Scientist Training Program, Chief, Cell Biology Branch, Division of Cell Biology and Biophysics, National Institute of General Medical Sciences (NIGMS), NIH
Shine, Kenneth I., M.D.	Executive Vice Chancellor for Health Affairs, University of Texas; President, Institute of Medicine (1992–2002)
Siegel, Richard M., M.D., Ph.D.	Investigator, Autoimmunity Branch, National Institute of Arthritis, Musculoskeletal, and Skin Diseases (NIAMS); Director, Trans-NIH Intramural M.D./Ph.D. Partnership Program, NIH
Sieving, Paul A., M.D., Ph.D.	Director, National Eye Institute (NEI), NIH
Singer, Daniel E., M.D.	Associate Dean for Clinical Programs, HMS, Boston, MA
Smith, Kent A.	Deputy Director, National Library of Medicine (NLM), NIH (1980–2004)
Smits, Helen L., M.D.	Chair, "Revitalizing the NIH Clinical Center for Tomorrow's Challenges" (issued 1996), NIH; Deputy Administrator, Health Care Financing Administration, DHHS (1993–1996)
Smolonsky, Marc	Director, Office of Legislative Policy and Analysis, NIH
Spaeth, Jennifer S.	Director, Office of Federal Advisory Committee Policy, NIH

Star, Robert A., M.D.	Director, Kidney, Urologic, and Hematologic Diseases, and Chief, Renal Diagnostics and Therapeutics Unit, National Institute of Diabetes and Digestive and Kidney Diseases (NIDDK), NIH
Strickland, Stephen P., Ph.D.	Senior Liaison Officer, Global Polio Eradication Initiative, UN Foundation, Washington, DC
Tabak, Lawrence A., D.D.S., Ph.D.	Director, National Institute of Dental and Craniofacial Research (NIDCR), NIH
Tacket, Carol O., M.D.	Professor of Medicine and Program Director, General Clinical Research Center, UMSOM, Baltimore, MD
Taffet, Richard ("Richie") M.	Office of Research Services (ORS), NIH
Thier, Samuel O., M.D.	Clinical Associate, National Institute of Arthritis and Metabolic Diseases, NIH (1962–1964); Professor of Medicine and Health Care Policy Emeritus, HMS, Boston, MA
Tilghman, Shirley M., Ph.D.	President, Princeton University, Princeton, NJ; Fogarty International Fellow, NIH (1975–1977)
Topol, Eric J., M.D.	Director, Scripps Translational Science Institute; Chief Academic Officer, Scripps Health, La Jolla, CA
Vagelos, P. Roy, M.D.	Trainee and Investigator, National Heart, Lung, and Blood Institute (NHLBI), NIH (1955–1966). Former CEO and Chair, Merck and Co., Inc.
Varmus, Harold E., M.D.	Clinical Associate, National Institute of Arthritis and Metabolic Diseases, NIH (1968–1970); Director, NIH (1993–1999); President, Memorial Sloan-Kettering Cancer Center, New York, NY; Nobel Prize in Physiology or Medicine, 1989
Venter, J. Craig, Ph.D.	Various positions, National Institute of Neurological Disorders and Stroke (NINDS), NIH; Chairman, Board of Directors, J. Craig Venter Institute, Rockville, MD
Volkow, Nora D., M.D.	Director, National Institute on Drug Abuse (NIDA), NIH

(*continued*)

(Continued)

Von Eschenbach, Andrew C., M.D.	Director, National Cancer Institute (NCI) (2002–2006), NIH.
Weinberg, Meryl	President, National Health Council, Washington, DC
Weisfeldt, Myron L., M.D.	Director (Chairman), Department of Medicine, Johns Hopkins University School of Medicine, Baltimore, MD
Weisman, Jennifer L., Ph.D.	Science and Technology Policy Fellow of the American Association for the Advancement of Science, National Institute of Dental and Craniofacial Research (NIDCR), NIH
Weissmann, Gerald, M.D.	Professor Emeritus of Medicine and Director, Biotechnology Study Center, New York University School of Medicine, New York, NY; Editor-in-Chief, *The FASEB Journal*
White, Jeffrey D., M.D.	Director, Office of Cancer Complementary and Alternative Medicine, National Cancer Institute (NCI), NIH
Wildenthal, Kern, M.D., Ph.D.	President, University of Texas Southwestern Medical Center at Dallas, TX
Wilder, Elizabeth L., Ph.D.	Acting Associate Director, Office of Portfolio Analysis and Strategic Initiatives, NIH
Woolley, Mary	President, Research!America, Alexandria, VA
Wyatt, Richard G., M.D.	Executive Director, Office of Intramural Research, NIH
Yamamoto, Keith R., Ph.D.	Executive Vice-Dean, University of California, San Francisco School of Medicine, San Francisco, CA
Yamata, Tadataka, M.D.	President, Global Health Program, Gates Foundation, Seattle, WA
Zerhouni, Elias A., M.D.	Director, NIH (2002–2008)
Zuk, Dorit, Ph.D.	AAAS/NIH Science and Technology Policy Fellow, Office of Extramural Research, Office of the Director (OD), NIH

APPENDIX B: ACRONYMS

Here are the meanings of most of the acronyms used in the text. The NIH employs many more. For a full list, see the NIH website: http://grants.nih.gov/grants/acronym_list.htm.

Acronym	Meaning
ABCR	Advisory Board for Clinical Research
AIDS	Acquired immune deficiency syndrome
BSC	Board of Scientific Counselors
CAT	Computerized axial tomography
CC	Clinical Center
CDC	Centers for Disease Control and Prevention
CIT	Center for Information Technology
CRADA	Cooperative Research and Development Agreement
CSR	Center for Scientific Review
CTSA	Clinical and Translational Science Award
D.D.S.	Doctor of Dental Surgery
DCRT	Division of Computer Research and Technology
DEA	Director of Extramural Activities
DEAS	Division of Extramural Administrative Support
DHHS	Department of Health and Human Services
DPCPSI	Division of Program Coordination, Planning, and Strategic Initiatives
DRFR	Division of Research Facilities and Resources
FAES	Foundation for Advanced Education in the Sciences
FFRDC	Federally Funded Research and Development Center
FIC	Fogarty International Center
GCRC	General Clinical Research Center
GPP	Graduate Partnership Program
HIV	Human immunodeficiency virus

(*continued*)

(Continued)

Acronym	Meaning
HPV	Human papilloma virus
HR	Human Resources
I/C	Institutes and Centers
IRB	Institutional Review Board
IRG	Integrated Review Group
J.D.	Juris Doctor
M.D.	Doctor of Medicine
M.S.W.	Master of Social Work
MDR	Multiple drug resistant
MHS	Marine Hospital Service
MRI	Magnetic resonance imaging
NAMI	National Alliance on Mental Illness
NCBI	National Center for Biotechnology Information
NCCAM	National Center for Complementary and Alternative Medicine
NCI	National Cancer Institute
NCMHD	National Center on Minority Health and Health Disparities
NCRR	National Center for Research Resources
NDEP	National Diabetes Education program
NEI	National Eye Institute
NHGRI	National Human Genome Research Institute
NHLBI	National Heart, Lung, and Blood Institute
NIA	National Institute on Aging
NIAAA	National Institute on Alcohol Abuse and Alcoholism
NIAID	National Institute of Allergy and Infectious Diseases
NIAMS	National Institute of Arthritis, Musculoskeletal, and Skin Diseases
NIBIB	National Institute of Biomedical Imaging and Bioengineering
NICHD	National Institute of Child Health and Human Development
NIDA	National Institute on Drug Abuse
NIDCD	National Institute on Deafness and Other Communication Disorders
NIDCR	National Institute of Dental and Craniofacial Research
NIDDK	National Institute of Diabetes and Digestive and Kidney Diseases
NIEHS	National Institute of Environmental Health Sciences
NIGMS	National Institute of General Medical Sciences
NIH	National Institutes of Health
NIMH	National Institute of Mental Health

NINDS	National Institute of Neurological Disorders and Stroke
NINR	National Institute of Nursing Research
NLM	National Library of Medicine
OD	Office of the Director
OHSR	Office of Human Subjects Research
OIHR	Office of International Health Research
OPASI	Office of Portfolio Analysis and Strategic Initiatives
ORS	Office of Research Services
ORWH	Office of Research on Women's Health
OTT	Office of Technology Transfer
Ph.D.	Doctor of Philosophy
PHS	Public Health Service
PI	Principal Investigator
PKD	Polycystic kidney disease
PMC	PubMed Central
R.N.	Registered Nurse
RFA	Request for applications
RFP	Request for proposals
SRA	Scientific Review Administrator

NOTES

Front Matter

1. Shalala, D.E. Miami, FL, 12/18/96, by e-mail, 11/15/07.
2. Ahrens, Jr., E.H. Quoted in *The crisis in clinical research: Overcoming institutional obstacles*. New York: Oxford University Press, 1992; p. 67.
3. Braunwald, Eugene, Boston, MA, by telephone, 3/2/07.
4. Porter, John E., Washington, DC, by telephone, 8/15/07.
5. Varmus, Harold, E. The director's small staff meeting.

Introduction

1. Wiltrout, Robert H., Bethesda, MD, by telephone, 7/8/08.
2. Potts, Jr., John T., Charlestown, MA, by telephone, 12/8/08.
3. Zerhouni, Elias A., Bethesda, MD, by e-mail, 1/14/09.

Chapter 1: Intramural Research Program

1. Gottesman, Michael M., Bethesda, MD, 2/9/07.
2. Wiltrout, Robert H., Bethesda, MD, by telephone, 7/8/08.
3. Korn, David, Washington, DC, by e-mail, 4/28/08.
4. Braunwald, Eugene, Boston, MA, by telephone, 3/2/07.
5. Pardes, Herbert, New York, NY, by telephone, 3/7/07.
6. Brown, Michael S., Wading River, NY, by telephone, 6/23/09.
7. Korn, David, Washington, DC, by telephone, 4/4/07.
8. Relman, Arnold S., Boston, MA, by telephone, 2/19/07.
9. Collins, Michael T., Bethesda, MD, 5/11/07.
10. Collins, Francis S., Bethesda, MD, 2/21/07.
11. Fauci, Anthony S., Bethesda, MD, 2/10/07.
12. Finkel, Toren, Bethesda, MD, by telephone, 10/2/07.
13. Gallo, Robert C., Baltimore, MD, 10/12/07.

14. Grady, Patricia A., Bethesda, MD, by e-mail, 9/4/07.
15. Lenardo, Michael J., Bethesda, MD, by telephone, 5/16/07.
16. Martin, Joseph B., Boston, MA, by telephone, 3/5/07.
17. Notkins, Abner L., Bethesda, MD, by telephone, 3/6/07.
18. Read, Elizabeth J., San Francisco, CA, by telephone, 5/23/07.
19. Schimpff, Stephen C., Baltimore, MD, 2/19/07.
20. Schwartz, David A., Research Triangle Park, NC, by telephone, 2/27/07.
21. Martensen, Robert L., Bethesda, MD, by e-mail, 7/17/08.
22. Lauer, Michael S., Bethesda, MD, by telephone, 3/7/08.
23. Caplan, Arthur L., Philadelphia, PA, by telephone, 3/21/07.
24. Hall, Zach W., Wilson, WY, by telephone, 12/4/07.
25. Jaffe, Holli Beckerman, Bethesda, MD, 5/24/07.
26. Gottesman, Michael M., Bethesda, MD, by e-mail, 10/15/08.
27. Zerhouni, Elias A., Bethesda, MD, by e-mail, 1/14/09.
28. Fraser-Liggett, Claire M., Baltimore, MD, 10/15/07.
29. Masur, Henry, Bethesda, MD, 4/23/07.
30. Pizzo, Philip A., Stanford, CA, by telephone, 7/13/07.
31. Kastor, J.A. The new medical school. In: *Specialty care in the era of managed care. Cleveland Clinic versus University Hospitals of Cleveland.* Baltimore: Johns Hopkins University Press, 2005, pp. 197–198.
32. Gottesman, Michael M., Bethesda, MD, by e-mail, 12/5/07.
33. Intramural professional designations and procedures. Available at http://www1.od.nih.gov/oir/sourcebook/prof-desig/prof-desig-toc.htm. 2007.
34. Gottesman, Michael M., Bethesda, MD, speech, 7/24/08.
35. Gottesman, Michael M., Bethesda, MD, by e-mail, 2/14/09.
36. Gottesman, Michael M., Bethesda, MD, by telephone, 8/26/08.
37. Garcia-Perez, Arlyn, Bethesda, MD, by e-mail, 11/1/07.
38. James, Stephen P., Bethesda, MD, by telephone, 2/23/08.
39. James, Stephen P., Bethesda, MD, by telephone, 10/29/07.
40. McKhann, Guy M., Baltimore, MD, by telephone, 6/15/07.
41. Kornberg, A. *The golden helix. Inside biotech ventures.* Paperback edition. Sausalito, CA: University Science Books, 2002; p. 24.
42. Park, B.S. The development of the intramural research program at the National Institutes of Health after World War II. *Perspectives in Biology and Medicine* 2003;46:388.
43. Altman, Lawrence, K. Arthur Kornberg, biochemist, dies at 89. *The New York Times*, 10/28/07.
44. Gottesman, Michael M., Bethesda, MD, by e-mail, 3/18/08.
45. Gallo, Robert C., Baltimore, MD, by e-mail, 10/8/07.
46. Kornberg, Arthur, Stanford, CA, speech 6/5/07.
47. Wyatt, Richard G., Bethesda, MD, 3/7/07.

48. Fauci, Anthony S., Bethesda, MD, by telephone, 1/8/08.
49. NIH Research Project Grant Program (R01). Available at http://grants1.nih.gov/grants/funding/r01.htm. 2007.
50. Scarpa, Antonio, Bethesda, MD, 6/22/07.
51. Landis, Story C., Bethesda, MD, by telephone, 1/9/08.
52. Kirschstein, Ruth L., Bethesda, MD, 2/8/07.
53. Research and training opportunities at the National Institutes of Health, 2007. Available at http://www.training.nih.gov/.
54. Milgram, Sharon L., Bethesda, MD, by telephone, 3/14/08.
55. Siegel, Richard M., Bethesda, MD, by telephone, 3/10/08.
56. Wyatt, Richard G., Bethesda, MD, by telephone, 9/15/08.
57. Gottesman, Michael M., Bethesda, MD, by telephone, 7/27/07.
58. Lenardo, Michael J., Bethesda, MD, by e-mail, 7/1/08.
59. Gallin, John I., Bethesda, MD, by e-mail, 11/13/07.
60. Oldfield, Edward H., Charlottesville, VA, by telephone, 11/5/07.
61. Klausner, Richard D., Seattle, WA, by telephone, 7/17/07.
62. Agnew, B. Scientists block NIH plan to grant Ph.D.s. *Science* 1999;284:1743.
63. Gallin, John I., Bethesda, MD, 4/23/07.
64. Harden, Victoria A., Bethesda, MD, by telephone, 3/5/07.
65. Tilghman, Shirley M., Princeton, NJ, by telephone, 8/10/07.
66. Varmus, Harold E., New York, NY, 3/17/07.
67. Kandel, E.R. *In search of memory: The emergence of a new science of mind.* New York: W.W. Norton, 2007.
68. DeVita, Jr., Vincent T., New Haven, CT, by telephone, 9/5/07.
69. Kirschner, Marc W., Boston, MA, by telephone, 7/17/07.
70. Marshall, E. Profile. Shirley Tilghman: Princeton's unconventional new chief. *Science* 2001;292:1288–1289.
71. Weissert, Will. NIH abandons proposal to create program in biomedical research. *Chronicle of Higher Education*, 9/24/99.
72. Varmus, Harold E., New York, NY, by e-mail, 9/6/08.
73. Foundation for Advanced Education in the Sciences. Available at http://www.faes.org/index.htm. 2007.
74. Noguchi, Constance Tom, Bethesda, MD, by telephone, 11/13/07.
75. Noguchi, Constance Tom, Bethesda, MD, by e-mail, 11/26/07.
76. Driefus, Claudia. Conversation with Robert L Martensen. A front-row seat as a health system goes awry. *The New York Times*, 1/20/09.
77. Taffet, Richard M., Bethesda, MD, by telephone, 4/7/08.
78. Public Health Service Commissioned Corps agencies and programs. file:///K:/NIH%20miscellaneous/Commissioned%20Corps%20PHS/Agencies%20and%20programs.htm. 2007.
79. Ayres, Elaine J., Bethesda, MD, by telephone, 5/18/07.
80. Hastings, Clare, Bethesda, MD, 4/27/07.

81. Wyatt, Richard G., Bethesda, MD, by e-mail, 5/21/07.
82. Levine, Arthur S., Pittsburgh, PA, by telephone, 7/16/07.
83. Klein, MK. The legacy of the "yellow berets." The Vietnam war, the doctor draft, and the NIH associate training program. Accessible at http://history.nih.gov/articles/YellowBerets.pdf. 1998.
84. U.S. Public Health Service Commissioned Corps, 2008. Available at http://www.usphs.gov/AboutUs/.
85. Park, B.S. The development of the intramural research program at the National Institutes of Health after World War II. *Perspectives in Biology and Medicine* 2003;46:334–336.
86. Special Forces, 2007. Available at http://banner.goarmy.com/banrtrck/banrdocs/armyop49.jsp;jsessionid=9E230E8BB973F4CD9AC7F963D0605C7A?banner=s002-sfrs-sf01-1003-49.
87. Potts, Jr., John T., Charlestown, MA, by telephone, 12/8/08.
88. Brown, Michael S., Wading River, NY, by e-mail, 6/26/08.
89. Broder, Samuel, Rockville, MD, by telephone, 7/27/07.
90. Klein, Harvey G., Bethesda, MD, by e-mail, 4/21/08.
91. Goldman, Lee, New York, NY, by telephone, 3/8/07.
92. Insel, Thomas R., Bethesda, MD, by telephone, 11/7/07.
93. Pizzo, Philip A., Stanford, CA, by telephone, 9/14/07.
94. Kelley, William N., Santa Ana, CA, by telephone, 2/15/07.
95. Gorden, Phillip, Bethesda, MD, 5/24/07.
96. Baltimore, David, Pasadena, CA, by telephone, 8/3/07.
97. Benz, Jr., Edward J., Boston, MA, by telephone, 5/21/07.
98. Briggs, Josephine P., Chevy Chase, MD, by telephone, 8/9/07.
99. Emanuel, Ezekiel J., Bethesda, MD, 3/27/07.
100. Goldstein, Joseph L., Dallas, TX, by telephone, 3/29/07.
101. Hyman, Steven E., Cambridge, MA, by telephone, 3/26/07.
102. Ramm, Louise E., Bethesda, MD, by telephone, 11/6/07.
103. Roth, Carl A., Bethesda, MD, by telephone, 5/16/08.
104. Venter, J. Craig, Rockville, MD, by telephone, 8/14/07.
105. Cohen, J. Is NIH's crown jewel losing luster? *Science* 1993;261: 1120–1127.
106. Cassell, Gail H., Indianapolis, IN, by telephone, 7/17/07.
107. Alter, Harvey J., Bethesda, MD, 4/23/07.
108. Klag, Michael J., Baltimore, MD, by telephone, 6/8/07.
109. Rosenberg, Steven A., Centerville, MA, by telephone, 8/14/07.
110. Shine, Kenneth I., Austin, TX, by telephone, 5/19/08.
111. Sieving, Paul A., Bethesda, MD, by telephone, 12/14/07.
112. Ehrenfeld, Ellie, Bethesda, MD, by telephone, 7/2/07.

113. Kaiser, J. U.S. biomedicine's mother ship braces for lab closings. *Science* 2008;319:1324–1325.
114. Charney, Dennis S., New York, NY, by telephone, 3/7/08.
115. Hollingsworth, J.R. Institutionalizing excellence in biomedical research: the case of the Rockefeller University. In: Stapleton, D.H., ed. *Creating a tradition of biomedical research: Contributions to the history of the Rockefeller University.* New York: Rockefeller University Press, 2004, Chap. 2.
116. Kaiser, Jocelyn, Washington, D.C., by telephone, 8/14/08.

Chapter 2: Extramural Research Program

1. Zerhouni, Elias A., Bethesda, MD, by e-mail, 1/14/09.
2. Martin, Joseph B., Boston, MA, by telephone, 3/5/07.
3. Marks, A.R. Rescuing the NIH before it is too late. *Journal of Clinical Investigation* 2006;116:844.
4. Harden, Victoria A., Bethesda, MD, by telephone, 3/5/07.
5. Marks, Andrew R., New York, NY, by telephone, 3/15/07.
6. Califf, Robert M., Durham, NC, by telephone, 4/2/07.
7. Fischbach, Gerald D., New York, NY, by telephone, 3/8/07.
8. Korn, D., Rich, R.R., Garrison, H.H., Golub, S.H., et al. Science policy. The NIH budget in the "postdoubling" era. *Science* 2002;296:1401–1402.
9. Alexander, D.F. Response to: Rescuing the NIH before it is too late. *Journal of Clinical Investigation* 2006;116:1462–1463.
10. Marks, A.R. Rescuing the NIH: the response. *Journal of Clinical Investigation* 2006;116:1460–1461.
11. Braunwald, Eugene, Boston, MA, by telephone, 3/2/07.
12. Goldman, Lee, New York, NY, by telephone, 3/8/07.
13. Weissmann, Gerald, New York, NY, by telephone, 3/15/07.
14. Weissmann, G. Planning science (a generation after Lewis Thomas). *Journal of Clinical Investigation* (Letter) 2006;116:1463.
15. Wadman, M. The map man. *Nature* 2007;448:406–407.
16. Corn, Milton, Bethesda, MD, by telephone, 10/30/07.
17. Ruiz Bravo, N. Response to: "Rescuing the NIH before it is too late" from the deputy director for extramural research. *Journal of Clinical Investigation* (Letter) 2006;116:1463–1464.
18. Ruiz Bravo, Norka, Bethesda, MD, 3/6/07.
19. An introduction to extramural research at the NIH. Available at http://grants.nih.gov/grants/intro2oer.htm. 2007.
20. Schaffer, Walter T., Bethesda, MD, by telephone, 9/9/08.
21. Ruiz Bravo, Norka, Bethesda, MD, by telephone, 6/12/08.
22. Varmus, Harold E., New York, NY, 1/18/08.
23. Scarpa, Antonio, Bethesda, MD, 6/22/07.

24. Landis, Story C., Bethesda, MD, by telephone, 1/9/08.
25. Yamamoto, Keith R., San Francisco, CA, by telephone, 3/4/08.
26. Collins, Michael T., Bethesda, MD, 5/11/07.
27. Nathan, D.G., Schechter, A.N. NIH support for basic and clinical research: Biomedical researcher angst in 2006. *Journal of the American Medical Association* 2006;295:2656–2658.
28. Gerber, Lynn H., Bethesda, MD, by telephone, 5/23/07.
29. DeVita, Jr., Vincent T., New Haven, CT, by telephone, 9/5/07.
30. Mahmoud, Adel, Princeton, NJ, by telephone, 7/6/07.
31. Martin, M.R., Lindquist, T., Kotchen, T. Why are peer review outcomes less favorable for clinical science than for basic science grant applications? *American Journal of Medicine* 2008;121:637–641.
32. Dickler, H.B., Fang, D., Heinig, S.J., Johnson, E., et al. New physician-investigators receiving National Institutes of Health research project grants: a historical perspective on the "endangered species." *Journal of the American Medical Association* 2007;297:2496–2501.
33. Shine, Kenneth I., Austin, TX, by telephone, 5/19/08.
34. Brown, Michael S., Wading River, NY, by telephone, 6/23/09.
35. Read, Elizabeth J., San Francisco, CA, by telephone, 5/23/07.
36. Briggs, Josephine P., Chevy Chase, MD, by telephone, 8/9/07.
37. Rodgers, Griffin P., Bethesda, MD, 9/13/07.
38. Klein, Harvey G., Bethesda, MD, 5/2/07.
39. Schechter, Alan N., Bethesda, MD, 3/27/07.
40. Schimpff, Stephen C., Baltimore, MD, 2/19/07.
41. Vagelos, P.R., Galambos, L. Hot science in big government. In: *Medicine, science, and Merck.* New York: Cambridge University Press, 2004, pp. 37–57.
42. Vagelos, P. Roy, Bedminster, NJ, by telephone, 1/6/09.
43. Ahrens, Jr., E.H. *The crisis in clinical research. Overcoming institutional obstacles.* New York: Oxford University Press, 1992.
44. Culliton, B.J., D'Auria, J. The physician-scientist really is an endangered species. *Journal of Investigative Medicine* 1998;46:417–419.
45. Goldstein, J.L., Brown, M.S. The clinical investigator: Bewitched, bothered, and bewildered—but still beloved. *Journal of Clinical Investigation* 1997;99:2803–2812.
46. Rosenberg, L. Physician-scientists—Endangered and essential. *Science* 1999;283:331–332.
47. Wyngaarden, J.B. The clinical investigator as an endangered species. *New England Journal of Medicine* 1979;301:1254–1259.
48. McKhann, Guy M., Baltimore, MD, by telephone, 6/15/07.
49. Nathan, D.G. Clinical research: Perceptions, reality, and proposed solutions. National Institutes of Health Director's Panel on Clinical Research. *Journal of the American Medical Association* 1998;280:1427–1431.

50. Nathan, D.G., Wilson, J.D. Clinical research and the NIH—A report card. *New England Journal of Medicine* 2003;349:1860–1865.
51. Nathan, David G., Boston, MA, by telephone, 6/8/07.
52. Roth, Carl A., Bethesda, MD, by telephone, 5/16/08.
53. Varmus, Harold E., New York, NY, by e-mail, 9/6/08.
54. Nathan, David G., Boston, MA, by e-mail, 6/12/07.
55. Schechter, A.N., Perlman, R.L., Rettig, R.A. Why is revitalizing clinical research so important, yet so difficult? *Perspectives in Biology and Medicine* 2004;47:476–486.
56. Butler, D. Crossing the valley of death. *Nature* 2008;453:840–842.
57. Korn, David, Washington, DC, by telephone, 4/4/07.
58. The Nobel Prize in Physiology or Medicine, 1989. Available at http://nobelprize.org/nobel_prizes/medicine/laureates/1989/press.html.
59. Bishop, J.M., Varmus, H., Bishop, J.M., Varmus, H. Re-aim blame for NIH's hard times. *Science* 2006;312:499.
60. Benz, Jr., Edward J., Boston, MA, by telephone, 5/21/07.
61. Thier, Samuel O., Boston, MA, by telephone, 3/1/07.

Chapter 3: Institutes

1. Croyle, Robert T., Rockville, MD, by telephone, 6/12/08.
2. McKhann, Guy M., Baltimore, MD, by telephone, 6/15/07.
3. Benz, Jr., Edward J., Boston, MA, by telephone, 5/21/07.
4. Gerber, Lynn H., Bethesda, MD, by telephone, 5/23/07.
5. Miller, Edward D., Baltimore, MD, 2/23/07.
6. Hall, Zach W., Wilson, WY, by telephone, 12/4/07.
7. Kirschstein, Ruth L., Bethesda, MD, 2/8/07.
8. Marshall, E. View from the top of a biomedical empire. Interview by Eliot Marshall. *Science* 1999;285:1654–1656.
9. Fraser-Liggett, Claire M., Baltimore, MD, 10/15/07.
10. Shreeve, J. *The genome war.* New York: Alfred, A. Knopf, 2004.
11. Wadman, M. High-profile departure ends genome institute's charmed run. 2007;13:518.
12. Collins, Francis S., Bethesda, MD, 2/21/07.
13. Fischbach, Gerald D., New York, NY, by telephone, 3/8/07.
14. Klausner, Richard D., Seattle, WA, by telephone, 7/9/07.
15. James, Stephen P., Bethesda, MD, by telephone, 10/29/07.
16. Rosenstein, Donald L., Bethesda, MD, by telephone, 11/8/07.
17. Demsey, Anthony, Bethesda, MD, 6/22/07.
18. Harden, Victoria A., Bethesda, MD, by telephone, 2/27/07.
19. Greenberger, Phyllis E., Washington, DC, by telephone, 7/17/07.

20. Varmus, H. Proliferation of National Institutes of Health. *Science* 2001;291: 1903–1905.
21. Check, E. Future of the NIH may lie in restructuring, committee told. *Nature* 2002;418:572.
22. Kaiser, J. Panel hears ideas for overhaul of NIH. *Science* 2002;297: 917–919.
23. Volkow, Nora D., Bethesda, MD, 2/8/07.
24. Enhancing the vitality of the National Institutes of Health: Organizational change to meet new challenges, 2003; p. 7. Available at http://books.nap.edu/catalog/10779.html.
25. Varmus, Harold E., New York, NY, by e-mail, 9/6/08.
26. Scarpa, Antonio, Bethesda, MD, 6/22/07.
27. Greenberg, D.S. *Science, money, and politics. Political triumph and ethical erosion.* Chicago: The University of Chicago Press, 2001; p. 475.
28. James, Stephen P., Bethesda, MD, by telephone, 8/1/07.
29. Spaeth, Jennifer S., Bethesda, MD, by telephone, 12/12/07.
30. Spaeth, Jennifer S., Bethesda, MD, by e-mail, 12/21/07.
31. James, Stephen P., Bethesda, MD, by telephone, 9/9/08, by e-mail, 8/19/08.
32. Enhancing the vitality of the National Institutes of Health: Organizational change to meet new challenges, 2003, page 11. Available at http://books.nap.edu/catalog/10779.html.
33. Selection criteria for NIH advisory committees, 2007. Available at http://www1.od.nih.gov/cmo/committee/SelectionCriteria2007.pdf.
34. Director's Council of Public Representatives, 2008. Available at http://copr.nih.gov/factsheet.asp.
35. Broder, Samuel, Rockville, MD, by telephone, 7/27/07.
36. Gottesman, Michael M., Bethesda, MD, 2/9/07.
37. National Cancer Act of 1971. National Cancer Advisory Board, 2007; p. 8. Available at http://www.cancer.gov/aboutnci/national-cancer-act-1971/.
38. President's Cancer Panel, 2008. Available at http://deainfo.nci.nih.gov/advisory/pcp/pcp.htm.
39. President's Cancer Panel. Biographies, 2008. Available at http://deainfo.nci.nih.gov/advisory/pcp/biographies.htm.
40. Niederhuber, John E., Bethesda, MD, 4/24/08.
41. National Cancer Institute. A–Z list of cancers, 2008. Available at http://www.cancer.gov/cancertopics/alphalist.
42. Folkers, Richard, Bethesda, MD, by e-mail, 9/19/08.
43. Gottesman, Michael M., Bethesda, MD, speech 7/24/08.
44. Reynolds, Craig W., Frederick, MD, by telephone, 5/29/08.
45. Klausner, Richard D., Seattle, WA, by e-mail, 4/21/08.
46. Kaiser, Jocelyn. Texas oncologist gets cancer institute post. *Science* 2001;294: 2263–2265.

47. Kirschstein, Ruth L., Bethesda, MD, by telephone, 5/15/08.
48. Greenwald, Harriet R., Bethesda, MD, by telephone, 2/7/08.
49. Driefus, Claudia. A conversation with Andrew von Eschenbach. Director tries to untangle web of cancer controversies. *The New York Times*, 3/11/03.
50. Committee on oversight and government reform. Breast cancer risks, 2003. Available at http://oversight.house.gov/features/politics_and_science/example_breast_cancer.htm.
51. Abortion and breast cancer. *The New York Times*, 1/6/03.
52. Abortion, miscarriage, and breast cancer risk. Accessed at file:///K:/NIH%20miscellaneous/Abortion%20and%20breast%20cancer%20NCI%20review.htm. 5/30/2003.
53. Nurse, P. U.S. biomedical research under siege. *Cell* 2006;124: 9–12.
54. Von Eschenbach, Andrew C., by telephone, 4/4/08.
55. Food and Drug Administration. Commissioners and their predecessors, 2008. Available at http://www.fda.gov/oc/commissioners/.
56. Niederhuber, J.E. A look inside the National Cancer Institute budget process: Implications for 2007 and beyond. *Cancer Research* 2007;67: 856–862.
57. Lenfant, Claude, 10/31/07, by telephone, Gaithersburg, MD.
58. Nabel, Elizabeth G., Bethesda, MD, 9/13/07.
59. Capell, B.C., Collins, F.S., Nabel, E.G. Mechanisms of cardiovascular disease in accelerated aging syndromes. *Circulation Research* 2007;101: 13–26.
60. Eriksson, M., Brown, W.T., Gordon, L.B., Glynn, M.W., et al. Recurrent *de novo* point mutations in lamin A cause Hutchinson-Gilford progeria syndrome. *Nature* 2003;423:293–298.
61. Cohn, Lawrence H., Boston, MA, by telephone, 10/17/07.
62. McIntosh, Charles L., Silver Spring, MD, by telephone, 12/12/07.
63. Waldhausen, J.A. The early history of congenital heart surgery: Closed heart operations. *Annals of Thoracic Surgery* 1997;64:1533–1539.
64. Finkel, Toren, Bethesda, MD, by telephone, 10/2/07.
65. McIntosh, Charles L., Silver Spring, MD, by e-mail, 4/22/08.
66. Passamani, Eugene R., Bethesda, MD, 10/3/07.
67. Baumgardner, William A., Baltimore, MD, by telephone, 8/8/07.
68. Weisfeldt, Myron L., Baltimore, MD, by telephone, 10/24/07.
69. Salganik, M. William. Suburban Hospital wins open-heart competition. *Baltimore Sun*, 12/11/02.
70. Kastor, J.A. MedStar Health. In: *Selling hospitals and practice plans: George Washington and Georgetown Universities*. Baltimore, MD: Johns Hopkins University Press, 2008, chap. 9.
71. Heart center plan rejected by high court. *Baltimore Sun*, 6/19/03.

72. Gragnolati, Brian A., Bethesda, MD, 10/3/07.
73. Susan Levine. Suburban's new NIH heart center offers more cardiac care options. *The Washington Post*, 9/30/06.
74. Horvath, Keith A., Bethesda, MD, 10/3/07.
75. Arai, Andrew E., Bethesda, MD, by telephone, 10/10/07.
76. Braunwald, Eugene, Boston, MA, by telephone, 3/2/07.
76a. Miranda S. Spivack. Hopkins Acquires Suburban Hospital. *The Washington Post*, 4/25/09.
77. Fauci, Anthony S., Bethesda, MD, by telephone, 8/9/07.
78. Centers for Disease Control (CDC). Kaposi's sarcoma and pneumocystis pneumonia among homosexual men—New York City and California. *Morbidity and Mortality Weekly Report* 1981;30:305–308.
79. Gallo, Robert C., by e-mail, 10/8/07.
80. Fauci, A.S. The syndrome of Kaposi's sarcoma and opportunistic infections: An epidemiologically restricted disorder of immunoregulation. *Annals of Internal Medicine* 1982;96:777–779.
81. Masur, Henry, Bethesda, MD, 4/23/07.
82. Henderson, D.K. Post-exposure treatment of HIV—taking some risks for safety's sake. *New England Journal of Medicine* 1997;337:1542–1543.
83. Henderson, David K., Bethesda, MD, 5/2/07.
84. Fauci, Anthony S., Bethesda, MD, by e-mail, 4/16/08.
85. Pizzo, Philip A., Stanford, CA, by telephone, 9/14/07.
86. Paul, William E., Bethesda, MD, by telephone, 11/11/07.
87. James, Stephen P., Bethesda, MD, by telephone, 7/3/07.
88. 2007 Lasker Awards. Available at http://www.laskerfoundation.org/awards/thisyear.html.
89. Brown, David M. AIDS vaccine testing at crossroads. After little progress in 25 years, scientists urge return to basic research. *The Washington Post*, 3/26/08.
90. Nabel, Gary J., Bethesda, MD, by telephone, 3/18/08.
91. Levine, Myron M., Baltimore, MD, 6/18/08.
92. Kaiser, J. Review of vaccine failure prompts a return to basics. *Science* 2008;320:30–31.
93. Tabak, Lawrence A., Bethesda, MD, 5/11/07.
94. Fradkin, Judith E., Bethesda, MD, by telephone, 9/23/08.
95. Star, Robert A., Bethesda, MD, by telephone, 10/8/08.
96. Goldman, Howard H., Potomac, MD, by telephone, 7/13/07.
97. Pardes, Herbert, New York, NY, by telephone, 3/7/07.
98. Insel, Thomas R., Bethesda, MD, by telephone, 6/17/08.
99. Leshner, Alan I., Washington, DC, by telephone, 8/28/08.
100. Hyman, Steven E., Cambridge, MA, by telephone, 3/26/07.
101. Goode, Erica. Scientist at work: Steve Hyman. Tireless, outspoken, and atypical, mental health chief rocks the boat. *The New York Times*, 6/15/99.

102. Insel, Thomas R., Bethesda, MD, by telephone, 11/7/07.
103. Rowland, L.P. *NINDS at 50*. New York: Demos Press, 2003.
104. National Institute of Neurological Disorders and Stroke, 2007. Available at http://www.nih.gov/about/almanac/organization/NINDS.htm.
105. Kirby, Kevin E., Houston, TX, by e-mail, 11/18/08.
106. Landis, Story C., Bethesda, MD, by telephone, 11/15/07.
107. Fischbach, Gerald D., New York, NY, by telephone, 8/24/07.
108. Rowland, Lewis P., New York, NY, by telephone, 2/19/08.
109. Landis, Story C., Bethesda, MD, by e-mail, 12/16/08.
110. Miles, W.D. *A history of the National Library of Medicine*. Bethesda, MD: National Institutes of Health, 1982.
111. Lindberg, Donald, A. B., Bethesda, MD, by telephone, 5/27/08.
112. Fee, Elizabeth, Bethesda, MD, by telephone, 6/10/08.
113. Fact sheet. The National Library of Medicine, 2008. Available at http://www.nlm.nih.gov/pubs/factsheets/nlm.html.
114. NIH Clinical Center. Facts at a glance, 2007. Available at http://clinicalcenter.nih.gov/about/welcome/fact.shtml.
115. Smith, Kent A., Bethesda, MD, by telephone, 9/24/08.
116. National Center for Biomedical Information, 2008. Available at http://www.ncbi.nlm.nih.gov/.
117. Landsman, David, Bethesda, MD, by telephone, 9/19/08.
118. Reminder concerning grantee compliance with public access policy and related NIH monitoring activities, 2008. Available at http://grants.nih.gov/grants/guide/notice-files/NOT-OD-08-119.html. 9/23/.
119. Lipman, David J., Bethesda, MD, by telephone, 9/24/08.
120. Alexander, Duane F., Bethesda, MD, 7/6/07.
121. Levine, Arthur S., Pittsburgh, PA, by telephone, 7/16/07.
122. Levine, Arthur S., Pittsburgh, PA, by e-mail, 5/13/08.
123. Alexander, Duane F., Bethesda, MD, by telephone, 10/7/08.
124. What makes this study important? 2008. http://www.nationalchildrensstudy.gov/about/overview/Pages/unique.aspx.
125. Kastor, J.A. Johns Hopkins University and Hospital: Unified governance. In: *Governance of teaching hospitals. Turmoil at Penn and Hopkins*. Baltimore: Johns Hopkins University Press, 2004; p. 274.
126. Berg, Jeremy M., Bethesda, MD, 3/27/07.
127. CRS report for Congress, 10/19/2006. Available at http://www.nih.gov/about/director/crsrept.pdf.
128. Park, B.S. The development of the intramural research program at the National Institutes of Health after World War II. *Perspectives in Biology and Medicine* 2003;46:396–399.
129. Kirschstein, Ruth L., Bethesda, MD, by telephone, 1/10/08.
130. Shapiro, Bert I., Bethesda, MD, by telephone, 8/14/08.

131. Rogers, Terry B., Baltimore, MD, by telephone, 9/3/08.
132. Link, A.N. *A generosity of spirit: The early history of the Research Triangle Park.* Research Triangle Foundation of North Carolina, 1995; p. 88.
133. Schwartz, David A., Research Triangle Park, NC, by telephone, 2/27/07.
134. Rosenwald, Michael S., Weiss, Rick. New ethics rules cost NIH another top researcher. *The Washington Post*, 4/2/05.
135. Weiss, Rick. Nominee for NIEHS has second thoughts. *The Washington Post*, 3/30/05.
136. Schwartz, David A., Research Triangle Park, NC, by e-mail, 5/28/08.
137. Kaiser, Jocelyn. Super-sized lab draws fire at NIH's environmental institute. *Science*, 6/7/07.
138. Weiss, Rick. 44 violated ethics rules, NIH director tells panel. *The Washington Post*, 7/15/05.
139. Grassley on NIEHS, 4/18/2008. Available at http://finance.senate.gov/press/Gpress/2008/prg041508.pdf.
140. Management review. National Institute of Environmental Health Sciences, 2008. Available at http://finance.senate.gov/press/Gpress/2008/prg041508a.pdf.
141. Schwartz, David A., Research Triangle Park, NC, by telephone, 3/18/08.
142. Kirschstein, Ruth L., Bethesda, MD, by telephone, 2/13/08.
143. National Institute on Alcohol Abuse and Alcoholism, 2007. Available at http://www.nih.gov/about/almanac/organization/NIAAA.htm.
144. Briggs, Josephine P., Chevy Chase, MD, by telephone, 8/9/07.
145. NIAAA budget, 2008. Available at http://www.niaaa.nih.gov/AboutNIAAA/CongressionalInformation/Budget.
146. Butler, Robert N., New York, NY, by telephone, 12/18/07.
147. Hodes, Richard J., Bethesda, MD, by telephone, 10/30/07.
148. Katz, Stephen I., Bethesda, MD, 7/5/07.
149. Lauer, Michael S., Bethesda, MD, by telephone, 3/7/08.
150. Grady, Patricia A., Bethesda, MD, by e-mail, 9/4/07.
151. Battey, Jr., James F., Bethesda, MD, by telephone, 2/12/08.
152. Battey, Jr., James F., Bethesda, MD, 5/11/07.
153. Varmus, Harold E., New York, NY, by e-mail, 2/18/08.
154. Kolberg, Rebecca, Bethesda, MD, by telephone, 8/1/08.
155. Schmeck, Harold M. Nobel winner to head gene project. *The New York Times*, 9/17/88.
156. Anderson, C. Watson resigns, genome project open to change. *Nature* 1992;356:549.
157. Healy, Bernadine P., Washington, DC, by telephone, 4/9/07.
158. DNA pioneer quits gene map project. *The New York Times*, 4/11/92.

159. Roberts, L. Why Watson quit as project head. *Science* 1992;256: 301–302.
160. Culliton, Barbara J., Washington, DC, by telephone, 8/4/07.
161. Lander, Eric S., Cambridge, MA, by telephone, 1/15/08.
162. Baltimore, David, Pasadena, CA, by telephone, 8/3/07.
163. Kaiser, J. Departing U.S. genome institute director takes stock of personalized medicine. *Science* 2008;320:1272.
164. Resignation of Francis Collins as director of the National Human Genome Research Institute, 5/28/2008. Available at http://www.genome.gov/Pages/About/OD/NewsandFeatures/FCOpeningRemarks.pdf.
165. Baum, Stanley, Philadelphia, PA, by telephone, 7/12/07.

Chapter 4: Centers

1. Welcome to Center for Scientific Review, 2007. Available at http://cms.csr.nih.gov/AboutCSR/Welcome+to+CSR/.
2. Fisher, Suzanne E., Bethesda, MD, 6/22/07.
3. The peer review process, 2007. Available at http://cms.csr.nih.gov/AboutCSR/OverviewofPeerReviewProcess.htm.
4. (deleted in press).
5. Scarpa, Antonio, Bethesda, MD, 6/22/07.
6. Scarpa, Antonio, Bethesda, MD, by e-mail, 2/6/08.
7. How scientists are selected for study section service, 2007. Available at http://cms.csr.nih.gov/PeerReviewMeetings/BestPractices/How+Scientists+Are+Selected+For+Study+Section+Service.htm.
8. Guidelines for study section chairs, 2007. Available at http://cms.csr.nih.gov/PeerReviewMeetings/BestPractices/Guidelines+for+Study+Section+Chairs.htm.
9. Charney, Dennis S., New York, NY, by telephone, 3/7/08.
10. Scarpa, Antonio, Europe, by e-mail, 7/11/07.
11. James, Stephen P., Bethesda, MD, by telephone, 8/1/07.
12. Demsey, Anthony, Bethesda, MD, 6/22/07.
13. Kirschner, Marc W., Boston, MA, by telephone, 8/3/07.
14. Hyman, Steven E., Cambridge, MA, by telephone, 3/26/07.
15. Yamamoto, Keith R., San Francisco, CA, by telephone, 3/4/08.
16. Brown, Michael S., Wading River, NY, by telephone, 6/23/09.
17. Landis, Story C., Bethesda, MD, by telephone, 1/9/08.
18. Nabel, Gary J., Bethesda, MD, by telephone, 3/18/08.
19. Roth, Carl A., Bethesda, MD, by telephone, 5/16/08.
20. Varmus, Harold E., New York, NY, 1/18/08.
21. Lauer, Michael S., Bethesda, MD, by telephone, 3/7/08.

22. Croyle, Robert T., Rockville, MD, by telephone, 6/12/08.
23. Mahmoud, Adel, Princeton, NJ, by telephone, 7/6/07.
24. McKhann, Guy M., Baltimore, MD, by telephone, 6/15/07.
25. Notkins, A.L. Erosion of freedom of inquiry. The NIH catalyst. 2006; July–August: 14–15.
26. Sieving, Paul A., Bethesda, MD, by telephone, 12/14/07.
27. Klausner, Richard D., Seattle, WA, by telephone, 7/17/07.
28. DeVita, Jr., Vincent T., New Haven, CT, by telephone, 9/5/07.
29. National Institute of Child Health and Human Development, 2007. Available at http://www.nichd.nih.gov/.
30. Baltimore, David, Pasadena, CA, by telephone, 8/3/07.
31. Haseltine, Florence P., Bethesda, MD.
32. Shamoo, A.E., Resnick, D.B. *Responsible conduct of research*. New York: Oxford University Press, 2003.
33. Shamoo, Adil E., Baltimore, MD, 2/16/07.
34. Ruiz Bravo, Norka, Bethesda, MD, by telephone, 6/12/08.
35. Fauci, Anthony S., Bethesda, MD, 2/10/07.
36. James, Stephen P., Bethesda, MD, by telephone, 7/3/07.
37. James, Stephen P., Bethesda, MD, by telephone, 2/23/08.
38. Grassley on NIEHS. Available at http://finance.senate.gov/press/Gpress/2008/prg041508.pdf. 4/18/2008.
39. Barros, Colleen F., Bethesda, MD, by telephone, 11/14/07.
40. Collins, Michael T., Bethesda, MD, 5/11/07.
41. Ramm, Louise E., Bethesda, MD, by telephone, 11/6/07.
42. James, Stephen P., Bethesda, MD, by telephone, 9/9/08; by e-mail, 8/19/08.
43. Katz, Stephen I., Bethesda, MD, 7/5/07.
44. NIH establishes working groups to examine peer review, 2007. Available at http://www.nih.gov/news/pr/jun2007/od-08.htm.
45. Reviewing peer review: NIH needs your help, 2007. Available at file:///k:/NIH%20miscellaneous/Peer%20Review%20Zerhouni.htm.
46. Yamamoto, Keith R., San Francisco, CA, by telephone, 8/24/08.
47. 2007–2008 peer review self study. Available at http://enhancing-peer-review.nih.gov/meetings/NIHPeerReviewReportFINALDRAFT.PDF. 2/29/2008.
48. Tabak, Lawrence A., Bethesda, MD, by telephone, 8/6/08.
49. 2007–2008 peer review self study. Available at http://enhancing-peer-review.nih.gov/meetings/NIHPeerReviewReport FINALDRAFT.PDF. 8/4/2008.
50. Kaiser, J. Changes in peer review target young scientists, heavyweights. *Science* 2008;320:1404.
51. Lederhendler, Israel I., Bethesda, MD, by telephone, 7/18/07.
52. Luckett, Donald N., Bethesda, MD, by e-mail, 10/27/08.

53. Scarpa, Antonio, Europe, by e-mail, 7/11/07.
54. Boyce, Thomas M., Bethesda, MD, by telephone, 7/9/07.
55. Electronic Research Administration. National Institutes of Health, 2007. Available at http://era.nih.gov/aboutera/index.cfm.
56. NIH Clinical Center, 2007. Available at http://clinicalcenter.nih.gov/.
57. Gallin, John I., Bethesda, MD, 4/23/07.
58. Greenberg, D.S. *Science, money, and politics. Political triumph and ethical erosion.* Chicago: The University of Chicago Press, 2001; pp. 193–194.
59. NIH Clinical Center. Facts at a glance, 2007. Available at http://clinicalcenter.nih.gov/about/welcome/fact.shtml.
60. Gallin, J.I., Varmus, H. Revitalization of the Warren G. Magnuson Clinical Center at the National Institutes of Health. *Academic Medicine* 1998;73:460–466.
61. Ayres, Elaine J., Bethesda, MD, by telephone, 5/18/07.
62. Rosenstein, Donald L., Bethesda, MD, by telephone, 11/8/07.
63. Gallin, John I., Bethesda, MD, by e-mail, 4/15/08.
64. Henderson, David K., Bethesda, MD, 5/2/07.
65. Byars, Sara, Bethesda, MD, by e-mail, 6/25/07.
66. Klein, Harvey G., Bethesda, MD, 5/2/07.
67. Hastings, Clare, Bethesda, MD, 4/27/07.
68. Wyatt, Richard G., Bethesda, MD, by e-mail, 5/21/07.
69. Pizzo, Philip A., Stanford, CA, by telephone, 7/13/07.
70. Klein, Harvey G., Bethesda, MD, by e-mail, 5/7/07.
71. Schimpff, Stephen C., Baltimore, MD, 2/19/07.
72. Patterson, Amy P., Bethesda, MD, by telephone, 4/4/08.
73. Menikoff, Jerry, Bethesda, MD, by telephone, 5/7/08.
74. Rosenberg, Steven A., Centerville, MA, by telephone, 8/14/07.
75. Smits, Helen L., Old Saybrook, CT, by telephone, 5/30/07.
76. Gallin, J.I., Smits, H.L. Managing the interface between medical schools, hospitals, and clinical research. *Journal of the American Medical Association* 1997;277:651–654.
77. Gormley, Maureen E., Bethesda, MD, 5/2/07.
78. Gallin, J.I. The need for clinical research education in the medical school curriculum. Proceedings of the Association of American Physicians. 1998;110:93–95.
79. Klag, Michael J., Baltimore, MD, by telephone, 6/8/07.
80. Fischbach, Gerald D., New York, NY, by telephone, 3/8/07.
81. Benz, Jr., Edward J., Boston, MA, by telephone, 5/21/07.
82. Coller, Barry S., New York, NY, by telephone, 5/10/07.
83. Advisory Board for Clinical Research. National Institutes of Health, 2007. Available at http://clinicalcenter.nih.gov/about/welcome/governance/advisory board.shtml.

84. NIH Clinical Center. Clinical research training, 2007. Available at http://clinicalcenter.nih.gov/researchers/training.shtml.
85. Gerber, Lynn H., Bethesda, MD, by telephone, 5/23/07.
86. Ognibene, Frederick P., Bethesda, MD, by telephone, 6/25/07.
87. Gallin, J.I., and Ognibene, F.P., eds. *Principles and practices of clinical research*. 2nd ed. New York: Academic Press, 2007.
88. Wilder, Elizabeth L., Bethesda, MD, 7/5/07.
89. Fleisher, Thomas A., Bethesda, MD, 5/2/07.
90. Alter, H.J., Houghton, M. Clinical Medical Research Award. Hepatitis C virus and eliminating post-transfusion hepatitis. *Nature Medicine* 2000;6:1082–1086.
91. Prestigious Lasker Award for 2000 honors Alter. *Clinical Center News* Oct. 2000; available at http://clinicalcenter.nih.gov/about/news/newsletter/2000/oct00/.
92. Alter, Harvey J., Bethesda, MD, 4/23/07.
93. Wiltrout, Robert H., Bethesda, MD, by telephone, 7/8/08.
94. Oldfield, Edward H., Charlottesville, VA, by telephone, 11/5/07.
95. Collins, Michael T., Bethesda, MD, by e-mail, 5/27/07.
96. Emanuel, E.J. Bioethics inside the beltway. The blossoming of bioethics at the NIH. *Kennedy Institute of Ethics Journal* 1998;8:455–466. Available at http://muse.jhu.edu/journals/kennedy_institute_of_ethics_journal/v008/8.4emanuel.html.
97. Emanuel, E.J. The NIH and bioethics: What should be done? *Academic Medicine* 2008;83:529–531.
98. Emanuel, Ezekiel J., Bethesda, MD, 3/27/07.
99. IDEA networks of biomedical research excellence, 2007. Available at http://www.ncrr.nih.gov/research_infrastructure/institutional_development_award/idea_networks_of_biomedical_research_excellence/.
100. Center for Information Technology. Available at http://www.nih.gov/about/almanac/organization/CIT.htm. 2008.
101. Graeff, Alan S., Bethesda, MD, by e-mail, 1/4/08.
102. Graeff, Alan S., Bethesda, MD, by telephone, 12/7/07.
103. Jones, John F., Bethesda, MD, by telephone, 2/20/08.
104. Glass, Roger I., Bethesda, MD, by telephone, 11/1/07.
105. Bridbord, Kenneth, Bethesda, MD, by telephone, 11/19/07.
106. Gottesman, Michael M., Bethesda, MD, speech 7/24/08.
107. Pathways to global health research. Strategic plan: 2008–2012 (2008). Available at http://www.fic.nih.gov/about/plan/strategicplan_08–12.htm.
108. National Center on Minority Health and Health Disparities, 2008. Available at http://www.nih.gov/about/almanac/organization/NCMHD.htm.
109. Berman, Brian M., Baltimore, MD, by telephone, 5/13/08.

110. Jacobs, Joseph J., Washington, DC, by telephone, 2/18/08.
111. Briggs, Josephine P., Bethesda, MD, by telephone, 8/25/08.
112. Briggs, Josephine P., Chevy Chase, MD, by telephone, 8/9/07.
113. Kirschstein, Ruth L., Bethesda, MD, by telephone, 11/8/07.
114. Eisenberg, D.M., Kessler, R.C., Foster, C., Norlock, F.E., et al. Unconventional medicine in the United States—prevalence, costs, and patterns of use. *New England Journal of Medicine* 1993;328:246–252.
115. Ashar, B.H., Rowland-Seymour, A. Advising patients who use dietary supplements. *American Journal of the Medical Sciences* 2008;121:91–97.
116. White, Jeffrey D., Bethesda, MD, by telephone, 3/6/08.
117. Briggs, Josephine P., Bethesda, MD, by telephone, 2/21/08.
118. Killen, John Y., Bethesda, MD, by telephone, 4/15/08.
119. Law, Catherine, Bethesda, MD, by telephone, 9/18/08.

Chapter 5: Finances

1. Notkins, Abner L., Bethesda, MD, by telephone, 3/6/07.
2. Masur, Henry, Bethesda, MD, 4/23/07.
3. Kelley, William N., Santa Ana, CA, by telephone, 2/15/07.
4. Varmus, Harold E., New York, NY, 3/17/07.
5. Harden, Victoria A., Bethesda, MD, by telephone, 2/27/07.
6. Greenberg, D.S. *Science, money, and politics. Political triumph and ethical erosion.* Chicago: The University of Chicago Press, 2001; p. 178.
7. Lee, Marvin, Bethesda, MD, by telephone, 2/1/07.
8. Shine, Kenneth I., Austin, TX, by telephone, 5/19/08.
9. Katz, Stephen I., Bethesda, MD, 7/5/07.
10. Greenberg, D.S. *Science, money, and politics. Political triumph and ethical erosion.* Chicago: The University of Chicago Press, 2001; p. 438.
11. Culliton, Barbara J., Washington, DC, by telephone, 8/4/07.
12. Korn, D., Rich, R.R., Garrison, H.H., Golub, S.H., et al. Science policy. The NIH budget in the "postdoubling" era. *Science* 2002;296: 1401–1402.
13. Alexander, D.F. Response to: Rescuing the NIH before it is too late. *Journal of Clinical Investigation* 2006;116:1462–1463.
14. Fauci, Anthony S., Bethesda, MD, by telephone, 1/8/08.
15. James, Stephen P., Bethesda, MD, by telephone, 8/1/07.
16. Altman, S., Bassler, B.L., Beckwith, J., Belfort, M., et al. An open letter to Elias Zerhouni. *Science* 2005;307:1409–1410.
17. DeVita, Jr., Vincent T., New Haven, CT, by telephone, 9/5/07.
18. Heinig, S.J., Krakower, J.Y., Dickler, H.B., Korn, D. Sustaining the engine of U.S. biomedical discovery. 2007;357:1042–1047.
19. Korn, David, Washington, DC, by telephone, 4/4/07.

20. Porter, John E., Washington, DC, by telephone, 8/15/07.
21. Henderson, David K., Bethesda, MD, 5/2/07.
22. Klein, Harvey G., Bethesda, MD, 5/2/07.
23. Baltimore, David, Pasadena, CA, by telephone, 8/3/07.
24. Brown, Michael S., Wading River, NY, by telephone, 6/23/09.
25. Goldman, Lee, New York, NY, by telephone, 3/8/07.
26. Harden, Victoria A., Bethesda, MD, by telephone, 3/5/07.
27. Miller, Edward D., Baltimore, MD, 2/23/07.
28. Nathan, David G., Boston, MA, by telephone, 6/8/07.
29. Niederhuber, John E., Bethesda, MD, 4/24/08.
30. Rubenstein, Arthur H., Philadelphia, PA, by telephone, 3/30/07.
31. Scarpa, Antonio, Bethesda, MD, 6/22/07.
32. Schechter, Alan N., Bethesda, MD, 3/27/07.
33. Mervis, J. U.S. science budget. NIH shrinks, NSF crawls as Congress finishes spending bills. *Science* 2006;311:28–29.
34. Germain, Ronald N., Bethesda, MD, by telephone, 8/9/07.
35. Ramm, Louise E., Bethesda, MD, by telephone, 11/6/07.
36. No way to run health research. *The New York Times*, 3/16/08.
37. Gottesman, Michael M., Bethesda, MD, by telephone, 8/26/08.
38. Schechter, Alan N., Bethesda, MD, by e-mail, 4/6/07.
39. Zerhouni, Elias A., Bethesda, MD, by e-mail, 1/14/09.
40. Wildenthal, Kern, Dallas, TX, by telephone, 8/16/07.
41. Levine, Arthur S., Pittsburgh, PA, by telephone, 7/16/07.
42. Schechter, Alan N., Bethesda, MD, by e-mail, 4/6/07.
43. Nurse, P. U.S. biomedical research under siege. *Cell* 2006;124:9–12.
44. Sieving, Paul A., Bethesda, MD, by telephone, 12/14/07.
45. Gibbons, Don L., Boston, MA, by telephone, 6/6/07.
46. Gorden, Phillip, Bethesda, MD, 5/24/07.
47. Braunwald, Eugene, Boston, MA, by telephone, 3/2/07.
48. Berg, Jeremy M., Bethesda, MD, 3/27/07.
49. Read, Elizabeth J., San Francisco, CA, by telephone, 5/23/07.
50. Pizzo, Philip A., Stanford, CA, by telephone, 7/13/07.
51. Brugge, Joan Siefert, Boston, MA, by e-mail, 6/21/07.
52. Fleisher, Thomas A., Bethesda, MD, 5/2/07.
53. Fleisher, Thomas A., Bethesda, MD, by e-mail, 4/14/08.
54. Mervis, J. 2007 U.S. budget. NIH trims award size as spending crunch looms. *Science* 2006;314:1862.
55. Couzin, J. Cancer research. Tight budget takes a toll on U.S.-funded clinical trials. *Science* 2007;315:1202–1203.
56. OMB circular A-76. Available at http://www.whitehouse.gov/omb/circulars/a076/a76_111402.doc. 2003.
57. Kelley, Rebecca L., Bethesda, MD, by telephone, 2/22/08.
58. Landis, Story C., Bethesda, MD, by telephone, 1/9/08.

59. Barros, Colleen F., Bethesda, MD, by telephone, 1/25/07.
60. Ruiz Bravo, Norka, Bethesda, MD, 3/6/07.
61. Roth, Carl A., Bethesda, MD, by telephone, 5/16/08.
62. Rohrbaugh, Mark L., Bethesda, MD, by telephone, 9/24/07.
63. CRADA, 2007. Available at http://www.ott.nih.gov/cradas/crada.html.
64. Foundation for the National Institutes of Health, 2007. Available at http://www.fnih.org/aboutus/What_Distinguishes_FNIH.shtml.
65. Porter, Amy McGuire, Bethesda, MD, by telephone, 9/6/07.
66. Sanders, Charles A., Chapel Hill, NC, by telephone, 9/10/07.
67. NIH Office of Technology Transfer activities, 2007. Available at http://www.ott.nih.gov/about_nih/statistics.html.
68. Ayres, Elaine J., Bethesda, MD, by telephone, 5/18/07.
69. Barros, Colleen F., Bethesda, MD, by telephone, 11/14/07.
70. Major, Christine M., Bethesda, MD, by telephone, 2/1/07.
71. Salary table 2008-dcb. Available at http://www.opm.gov/oca/08tables/html/dcb.asp. 2008.
72. IC director/NIH deputy director compensation model. 2005.
73. Gottesman, Michael M., Bethesda, MD, speech 7/24/08.
74. Zerhouni, Elias A., Pasadena, MD, by e-mail, 7/16/09.

Chapter 6: Congress and Advocates

1. Porter, John E., Washington, DC, by telephone, 8/15/07.
2. Porter, John E., Washington, DC, by e-mail, 4/25/08.
3. Kelley, William N., Santa Ana, CA, by telephone, 2/15/07.
4. Harris, Gardner Specter, a fulcrum of the stimulus bill, pulls off a coup for health money. *The New York Times*, 2/14/09.
5. Rogers, Paul G., Washington, DC, by telephone, 5/2/08.
6. Havesi, Dennis. Paul G. Rogers, "Mr. Health" in Congress, is dead at 87. *The New York Times*, 10/15/08.
7. Zerhouni, Elias A., Pasadena, MD, by e-mail, 7/16/09.
8. Culliton, Barbara J., Washington, DC, by telephone, 8/4/07.
9. Varmus, Harold E., New York, NY, 3/17/07.
10. Rogers, Paul G., Washington, DC, by telephone, 12/20/07.
11. Greenberger, Phyllis E., Washington, DC, by telephone, 7/17/07.
12. Weinberg, Myrl, Washington, DC, by telephone, 7/16/07.
13. Fischbach, Gerald D., New York, NY, by telephone, 3/8/07.
14. Fauci, Anthony S., Bethesda, MD, by telephone, 1/8/08.
15. Mahmoud, Adel, Princeton, NJ, by telephone, 7/6/07.
16. Hyman, Steven E., Cambridge, MA, by telephone, 3/26/07.
17. Research!America. An alliance for discoveries in health. Available at http://www.researchamerica.org/about. 2008.

18. Woolley, Mary, Alexandria, VA, by telephone, 4/3/07.
19. Gibbons, Don L., Boston, MA, by telephone, 6/6/07.
20. Baltimore, David, Pasadena, CA, by telephone, 8/3/07.
21. Miller, Edward D., Baltimore, MD, 2/23/07.
22. Casey, Kevin, Cambridge, MA, by telephone, 5/14/07.
23. Martin, Joseph B., Boston, MA, by telephone, 3/5/07.
24. Within our grasp or slipping away. Assuring a new era of scientific and medical progress. A statement by a group of concerned universities and research institutions (2007). Available at http://hms.harvard.edu/public/news/nih_funding.pdf.
25. Brugge, Joan Siefert. Testimony. 3/19/2007.
26. Ruiz Bravo, Norka, Bethesda, MD, 3/6/07.
27. Zerhouni, E.A. NIH in the post-doubling era: Realities and strategies. *Science* 2006;314:1088–1090.
28. Pizzo, Philip A., Stanford, CA, by telephone, 9/14/07.
29. Mervis, J. U.S. science adviser tells researchers to look elsewhere. *Science* 2007;316: 817–818.
30. Varmus, Harold E., New York, NY, 1/18/08.

Chapter 7: Directors

1. Butler, Robert N., New York, NY, by telephone, 12/18/07.
2. Harden, Victoria A., Bethesda, MD, by telephone, 2/27/07.
3. Park, B.S. The development of the intramural research program at the National Institutes of Health after World War II. *Perspectives in Biology and Medicine* 2003;46:389–394.
4. Kirschner, Marc W., Boston, MA, by telephone, 7/17/07.
5. Collins, Francis S., Bethesda, MD, 2/21/07.
6. Fauci, Anthony S., Bethesda, MD, 2/10/07.
7. Kelley, William N., Santa Ana, CA, by telephone, 2/15/07.
8. Miller, Edward D., Baltimore, MD, 2/23/07.
9. Fallows, James. The political scientist. *The New Yorker*, 6/7/99.
10. Broad, William J., Dean, Cornelia. Rivals' visions differ on unleashing innovation. *The New York Times*, 10/17/08.
11. Shine, Kenneth I., Austin, TX, by telephone, 5/19/08.
12. Alonso-Zaldivar, R., Kaplan, K. Loosening of stem cell limits backed. *Los Angeles Times*, 3/20/07.
13. Hyman, Steven E., Cambridge, MA, by telephone, 3/26/07.
14. Gorden, Phillip, Bethesda, MD, 5/24/07.
15. Enhancing the vitality of the National Institutes of Health: Organizational change to meet new challenges. Available at http://books.nap.edu/catalog/10779.html. 2003, page 11.

16. Rettig, R.A. Reorganizing the National Institutes of Health. *Health Affairs* 2004;23:257–262.
17. Relman, Arnold S., Boston, MA, by telephone, 2/19/07.
18. Enhancing the vitality of the National Institutes of Health: Organizational change to meet new challenges. Available at http://books.nap.edu/catalog/10779.html. 2003, page 9.
19. Office of the Director, 2008. Available at http://www.nih.gov/icd/od/offices.htm.
20. Varmus, Harold E., New York, NY, by e-mail, 9/6/08.
21. Fauci, Anthony S., Bethesda, MD, by e-mail, 7/16/09.
22. Kastor, J.A. The new medical school. In: *Specialty care in the era of managed care. Cleveland Clinic versus University Hospitals of Cleveland*. Baltimore: Johns Hopkins University Press, 2005, pp. 199–201.
23. Kastor, J.A. Cleveland Clinic: the clinical factory. In: *Specialty care in the era of managed care. Cleveland Clinic versus University Hospitals of Cleveland*. Baltimore: Johns Hopkins University Press, 2005, pp. 5–83.
24. Healy, Bernadine P., Washington, DC, by telephone, 4/9/07.
25. Schechter, Alan N., Bethesda, MD, 3/27/07.
26. Lee, Marvin, Bethesda, MD, by telephone, 2/1/07.
27. Cohen, J. Bernadine Healy bows out. *Science* 1993;259:1388–1389.
28. Varmus, Harold E., New York, NY, 3/17/07.
29. Gerber, Lynn H., Bethesda, MD, by telephone, 5/23/07.
30. Greenberger, Phyllis E., Washington, DC, by telephone, 7/17/07.
31. Haseltine, Florence P., Bethesda, MD.
32. Haseltine, F.P., ed. *Women's health research. A medical and policy primer.* Washington, DC: Health Press International, 1997.
33. Kirschstein, Ruth L., Bethesda, MD, 2/8/07.
34. Pinn, Vivian W., Bethesda, MD, by telephone, 8/21/07.
35. Office of Research on Women's Health, 2007. Available at http://orwh.od.nih.gov/about.html.
36. Klein, Harvey G., Bethesda, MD, 5/2/07.
37. Culliton, Barbara J., Washington, DC, by telephone, 8/4/07.
38. Greenberg, D.S. *Science, money, and politics. Political triumph and ethical erosion.* Chicago: The University of Chicago Press, 2001; pp. 222–223.
39. Rodgers, Griffin P., Bethesda, MD, 9/13/07.
40. Warren, E. Leary. U.S. scientists to seek patent on 2,375 more genes. *The New York Times*, 2/13/92.
41. DNA pioneer quits gene map project. *The New York Times*, 4/11/92.
42. Venter, J. Craig, Rockville, MD, by e-mail, 10/31/08.
43. Venter, J.C. A life decoded. *My genome: my life.* New York: Viking Press, 2008.
44. Greenberg, D.S. *Science, money, and politics. Political triumph and ethical erosion.* Chicago: The University of Chicago Press, 2001; p. 108.

45. Chen, Philip S., Ooltewah, TN, by telephone, 8/17/07.
46. Corn, Milton, Bethesda, MD, by telephone, 10/30/07.
47. Greenwald, Harriet R., Bethesda, MD, by telephone, 2/7/08.
48. Hall, Zach W., Wilson, WY, by telephone, 12/4/07.
49. Harden, Victoria A., Bethesda, MD, by telephone, 3/5/07.
50. Kirschstein, Ruth L., Bethesda, MD, by telephone, 4/10/07.
51. Klausner, Richard D., Seattle, WA, by telephone, 7/9/07.
52. Levine, Arthur S., Pittsburgh, PA, by telephone, 7/16/07.
53. Liotta, Lance A., Manassas, VA, by telephone, 6/18/07.
54. Mahmoud, Adel, Princeton, NJ, by telephone, 7/6/07.
55. Paul, William E., Bethesda, MD, by telephone, 11/11/07.
56. Pizzo, Philip A., Stanford, CA, by telephone, 7/13/07.
57. Ramm, Louise E., Bethesda, MD, by telephone, 11/6/07.
58. Roth, Carl A., Bethesda, MD, by telephone, 5/16/08.
59. Rubenstein, Arthur H., Philadelphia, PA, by telephone, 3/30/07.
60. Thier, Samuel O., Boston, MA, by telephone, 3/1/07.
61. Venter, J. Craig, Rockville, MD, by telephone, 8/14/07.
62. Wyatt, Richard G., Bethesda, MD, 3/7/07.
63. Yamada, Tadataka, Chicago, IL, by telephone, 4/13/07.
64. Zerhouni, Elias A., Bethesda, MD, 1/24/07.
65. Hodes, Richard J., Bethesda, MD, by telephone, 10/30/07.
66. Sontag, Deborah. Who brought Bernadine Healy down? *The New York Times*, 12/23/01.
67. Varmus, H. *The art and politics of science*. W. W. Norton and Company, 2009.
68. Dizikes, Peter. Political science. *The New York Times*, 2/15/09.
69. Leshner, Alan I., Washington, DC, by telephone, 8/28/08.
70. Marshall, E. View from the top of a biomedical empire. Interview by Eliot Marshall. *Science* 1999;285:1654–1656.
71. Fischbach, Gerald D., New York, NY, by telephone, 3/8/07.
72. McKhann, Guy M., Baltimore, MD, by telephone, 6/15/07.
73. Schechter, Alan N., Bethesda, MD, by telephone, 2/17/08.
74. Schechter, Alan N., Bethesda, MD, by e-mail, 3/17/08.
75. Wiltrout, Robert H., Bethesda, MD, by telephone, 7/8/08.
76. Grady, Patricia A., Bethesda, MD, by e-mail, 9/4/07.
77. Levine, Arthur S., Pittsburgh, PA, by e-mail, 5/13/08.
78. Varmus, H. Shattuck Lecture—Biomedical research enters the steady state. *New England Journal of Medicine* 1995;333:811–815.
79. The National Institutes of Health (NIH): Organization, funding and congressional issues. Available at http://www.nih.gov/about/director/crsrept.pdf. 10/19/2006.
80. Bishop, J.M., Kirschner, M., Varmus, H., Bishop, J.M., et al. Science and the new administration. *Science* 1993;259:444–445.

81. Lenardo, Michael J., Bethesda, MD, by telephone, 5/16/07.
82. Scientific Interest Groups, 2007. Available at http://www.nih.gov/sigs/.
83. Alter, Harvey J., Bethesda, MD, 4/23/07.
84. Katz, Stephen I., Bethesda, MD, 7/5/07.
85. Landis, Story C., Bethesda, MD, by telephone, 1/9/08.
86. Lenfant, Claude, Gaithersburg, MD, by telephone, 10/31/07.
87. Levine, Myron M., Baltimore, MD, 6/18/08.
88. Schechter, Alan N., Bethesda, MD, by e-mail, 4/6/07.
89. Harris, Gardner. Four top science advisors are named by Obama. *The New York Times*, 12/21/08.
90. Brainard, J. Long-term vacancy at the NIH. *The Chronicle of Higher Education*, 2/1/02.
91. Fischbach, Gerald D., New York, NY, by telephone, 8/24/07.
92. Zerhouni, Elias A., Pasadena, MD, by e-mail, 7/16/09.
93. Fauci, Anthony S., Bethesda, MD, by e-mail, 7/28/09.
94. Connolly, Ceci, Milbank, Dana. Bush selects a top official at Johns Hopkins to head NIH. *The Washington Post*, 3/6/02.
94a. Statement by HHS Secretary Kathleen Sebelius on the Passing of Dr. Ruth Kirschstein. NIH Press Release October 9, 2009.
95. Kaiser, J. U.S. appointment. Zerhouni confirmed as NIH director. *Science* 2002;296:997–999.
96. Klag, Michael J., Baltimore, MD, by telephone, 6/8/07.
97. Caplan, Arthur L., Philadelphia, PA, by telephone, 3/21/07.
98. Nathan, David G., Boston, MA, by telephone, 6/8/07.
99. Porter, John E., Washington, DC, by telephone, 8/15/07.
100. Krensky, Alan M., Bethesda, MD, by telephone, 8/17/07.
101. The director's message—NIH Reform Act of 2006 (2007). Available at http://www.nih.gov/about/reauthorization/reauthorizationmessage.html.
102. Varmus, H. Proliferation of National Institutes of Health. *Science* 2001;291:
1903–1905.
103. Alexander, D.F. Response to "Rescuing the NIH before it is too late." *Journal of Clinical Investigation* 2006;116:1462–1463.
104. Baltimore, David, Pasadena, CA, by telephone, 8/3/07.
105. Braunwald, Eugene, Boston, MA, by telephone, 3/2/07.
106. Briggs, Josephine P., Chevy Chase, MD, by telephone, 8/9/07.
107. Charney, Dennis S., New York, NY, by telephone, 3/7/08.
108. Coller, Barry S., New York, NY, by telephone, 5/10/07.
109. Croyle, Robert T., Rockville, MD, by telephone, 6/12/08.
110. Fauci, Anthony S., Bethesda, MD, by telephone, 1/8/08.
111. Finkel, Toren, Bethesda, MD, by telephone, 10/2/07.
112. Gormley, Maureen E., Bethesda, MD, 5/2/07.

113. Insel, Thomas R., Bethesda, MD, by telephone, 11/7/07.
114. James, Stephen P., Bethesda, MD, by telephone, 8/1/07.
115. Kleinman, Dushanka V., College Park, MD, by telephone, 7/16/07.
116. Kornberg, Arthur, Stanford, CA, speech 6/5/07.
117. Lauer, Michael S., Bethesda, MD, by telephone, 3/7/08.
118. Nabel, Gary J., Bethesda, MD, by telephone, 3/18/08.
119. Nathan, D.G., Schechter, A.N. NIH Support for basic and clinical research: Biomedical researcher angst in 2006. *Journal of the American Medical Association* 2006;295:2656–2658.
120. Niederhuber, John E., Bethesda, MD, 4/24/08.
121. Oldfield, Edward H., Charlottesville, VA, by telephone, 11/5/07.
122. Scarpa, Antonio, Bethesda, MD, 6/22/07.
123. Sieving, Paul A., Bethesda, MD, by telephone, 12/14/07.
124. Tacket, Carol O., Baltimore, MD, 5/16/07.
125. Wilder, Elizabeth L., Bethesda, MD, 7/5/07.
126. Zerhouni, Elias A., Pasadena, MD, by e-mail, 7/15/09.
127. Overview of the NIH Roadmap, 2007. Available at http://nihroadmap.nih.gov/overview.asp.
128. Brown, Michael S., Wading River, NY, by telephone, 6/23/09.
129. FY 2007 final enacted appropriation (2007). Available at http://officeofbudget.od.nih.gov/PDF/Final%20Conference%20by%20IC%20for%20Web.pdf.
130. Weissmann, Gerald, New York, NY, by telephone, 3/15/07.
131. Berg, Jeremy M., Bethesda, MD, 3/27/07.
132. Califf, Robert M., Durham, NC, by telephone, 4/2/07.
133. FitzGerald, Garret A., Philadelphia, PA, by telephone, 7/11/07.
134. Ford, Daniel E., Baltimore, MD, by telephone, 8/16/07.
135. Hayward, Anthony R., Bethesda, MD, by telephone, 8/6/07.
136. Kaiser, J. NIH funds a dozen "homes" for translational research. *Science* 2006;314:237.
137. Clinical and Translational Science Awards (Consortium), 2007. Available at file:///K:/NIH%20miscellaneous/CTSA.asp.
138. Tabak, Lawrence A., Bethesda, MD, 5/11/07.
139. Volkow, Nora D., Bethesda, MD, 2/8/07.
140. Pizzo, Philip A., Stanford, CA, by telephone, 9/14/07.
141. Zerhouni, Elias A., Bethesda, MD, by e-mail, 1/14/09.
142. Ruiz Bravo, Norka, Bethesda, MD, by e-mail, 6/25/08.
143. Ruiz Bravo, Norka, Bethesda, MD, 3/6/07.
144. Ognibene, Frederick P., Bethesda, MD, by telephone, 6/25/07.
145. Clinical and Translational Science Awards to transform clinical research, 2007. Available at file:///K:/NIH%20miscellaneous/CTSA%20first%2012%20awards.htm.
146. Singer, Daniel E., Boston, MA, by telephone, 6/18/07.

147. NIH Director's Pioneer Award, 2007. Available at http://nihroadmap.nih.gov/pioneer/.
148. Office of Portfolio Analysis and Strategic Initiatives, 2007. Available at http://opasi.nih.gov/about.asp.
149. Kaiser, J. Drawing a map for twenty-seven divisions in NIH's army. *Science* 2007;317:887.
150. Wadman, M. The map man. *Nature* 2007;448:406–407.
151. Biographical sketch of Dr. Elias A. Zerhouni, 2007. Available at http://www.nih.gov/about/director/directorbio.htm.
152. Curriculum vita of Elias A. Zerhouni, 2007. Available at file:///k:/NIH%20CVs/Zerhouni.htm.
153. Rogers, Paul G., Washington, DC, by telephone, 12/20/07.
154. Schwartz, David A., Research Triangle Park, NC, by telephone, 2/27/07.
155. Yamamoto, Keith R., San Francisco, CA, by telephone, 3/4/08.
156. Vagelos, P. Roy, Bedminster, NJ, by telephone, 1/6/09.
157. Harris, Gardner. Federal health official to step down. *The New York Times*, 9/25/08.
158. Harris, Gardner. Pick to lead health agency draws praise and some concern. *The New York Times*, 7/9/09.
158a. Senate confirms new director. *The Washington Post* 8/7/09
158b. Sam Harris. Science is in the Details. *The New York Times* 7/27/09
159. Fauci, Anthony S., Bethesda, MD, 10/30/08.

Chapter 8: Controversies

1. Caplan, Arthur L., Philadelphia, PA, by telephone, 3/21/07.
2. Culliton, Barbara J., Washington, DC, by telephone, 8/4/07.
3. Grady, Patricia A., Bethesda, MD, by e-mail, 9/4/07.
4. Alter, Harvey J., Bethesda, MD, 4/23/07.
5. Briggs, Josephine P., Chevy Chase, MD, by telephone, 8/9/07.
6. Collins, Michael T., Bethesda, MD, 5/11/07.
7. DeVita, Jr., Vincent T., New Haven, CT, by telephone, 9/5/07.
8. Emanuel, Ezekiel J., Bethesda, MD, 3/27/07.
9. Fauci, Anthony S., Bethesda, MD, by telephone, 8/9/07.
10. Greenberger, Phyllis E., Washington, DC, by telephone, 7/17/07.
11. Haseltine, Florence P., Bethesda, MD, 9/21/2007.
12. Jaffe, Holli Beckerman, Bethesda, MD, 5/24/07.
13. Klein, Harvey G., Bethesda, MD, 5/2/07.
14. Notkins, Abner L., Bethesda, MD, by telephone, 3/6/07.
15. Rodgers, Griffin P., Bethesda, MD, 9/13/07.
16. Schechter, Alan N., Bethesda, MD, 3/27/07.
17. Tabak, Lawrence A., Bethesda, MD, 5/11/07.

18. Venter, J. Craig, Rockville, MD, by telephone, 8/14/07.
19. Wiltrout, Robert H., Bethesda, MD, by telephone, 7/8/08.
20. Steinbrook, R. Financial conflicts of interest and the NIH. *New England Journal of Medicine* 2004;350:327–330.
21. Zerhouni, Elias A., Pasadena, MD, by e-mail, 7/16/09.
22. Fauci, Anthony S., Bethesda, MD, 2/10/07.
23. Healy, Bernadine P., Washington, DC, by telephone, 4/9/07.
24. Jaffe, Holli Beckerman, Bethesda, MD, by telephone, 6/19/07.
25. Battey, Jr., James F., Bethesda, MD, 5/11/07.
26. Rich, Eric. NIH scientist pleads guilty in accepting $285,000 from Pfizer. *The Washington Post*, 12/9/06.
27. Willman, David. NIH researcher is ordered to forfeit Pfizer payments. The scientist, who pleaded guilty to conflict of interest, gets two years' probation and community service. *Los Angeles Times*, 12/23/06.
28. Willman, David. Stealth merger: drug companies and government medical research. *Los Angeles Times*, 12/7/03.
29. Willman, David. The National Institutes of Health. Public servant or private marketer? *Los Angeles Times*, 12/22/04.
30. Fahrenthold, David A. U.S. criminal charges filed against scientist. Undisclosed consulting deals at issue. *The Washington Post*, 12/5/06.
31. Weiss, Rick. "Serious misconduct" by NIH expert found. *The Washington Post*, 6/14/06.
32. Alzheimer's disease: Advances and hope, 2003. Trey Sunderland, http://www.nih.gov/.
33. Insel, Thomas R., Bethesda, MD, by telephone, 11/7/07.
34. Weiss, Rick. NIH punishments criticized. Members of Congress complain that sanctions are too "soft." *The Washington Post*, 9/14/06.
35. (deleted in press).
36. Weiss, Rick. NIH officials investigated for possible ethics breach. *The Washington Post*, 6/27/03.
37. Willman, David. NIH audit criticizes scientist's dealings. *Los Angeles Times*, 9/10/06.
38. Rosenwald, Michael S., Weiss, Rick. New ethics rules cost NIH another top researcher. *The Washington Post*, 4/2/05.
39. Marshall, E. Conflict of interest. Zerhouni pledges review of NIH consulting in wake of allegations. *Science* 2003;302:2046.
40. Weiss, Rick. NIH chief taps duo to review allegations. *The Washington Post*, 1/23/04.
41. Report of the National Institutes of Health blue-ribbon panel on conflict of interest policies. Available at www.nih.gov/about/ethics_COI_panelreport.pdf. 6/22/2004.
42. Kaiser, J. Biomedical research: Feeling the heat, NIH tightens conflict-of-interest rules. *Science* 2004;305:25–26.

43. Weiss, Rick. NIH panel backs tighter limits on outside income. *The Washington Post*, 5/7/04.
44. Pizzo, Philip A., Stanford, CA, by telephone, 9/14/07.
45. Weiss, Rick. House panel scolds NIH chief, HHS. *The Washington Post*, 5/13/04.
46. Varmus, Harold E., New York, NY, 3/17/07.
47. Gorden, Phillip, Bethesda, MD, 5/24/07.
48. Weiss, Rick. NIH will restrict outside income. *The Washington Post*, 2/2/05.
49. Weiss, Rick. NIH workers angered by new ethics rules. Restrictions on outside income meet with derision at meeting. *The Washington Post*, 2/3/05.
50. Roth, Carl A., Bethesda, MD, by telephone, 5/16/08.
51. Gerber, Lynn H., Bethesda, MD, by telephone, 5/23/07.
52. Fleisher, Thomas A., Bethesda, MD, 5/2/07.
53. Harden, Victoria A., Bethesda, MD, by telephone, 3/5/07.
54. James, Stephen P., Bethesda, MD, by telephone, 7/3/07.
55. Steinbrook, R. Standards of ethics at the National Institutes of Health. *New England Journal of Medicine* 2005;352:1290–1292.
56. Weiss, Rick. NIH to set stiff restrictions on outside consulting. *The Washington Post*, 8/4/04.
57. Weiss, Rick. NIH bans collaboration with outside companies. Policy comes after conflict-of-interest inquiry. *The Washington Post*, 9/24/04.
58. Evaluation of the impact of the new NIH ethics rules on recruitment and retention. Available at http://www.nih.gov/about/ethics/evaluationslides.pdf. 2006.
59. Weiss, Rick. NIH clears most researchers in conflict-of-interest probe. *The Washington Post*, 2/23/05.
60. Gottesman, Michael M., Bethesda, MD, speech 7/24/08.
61. Weiss, Rick. 44 violated ethics rules, NIH director tells panel. *The Washington Post*, 7/15/05.
62. Willman, David. NIH inquiry shows widespread ethical lapses, lawmaker says. *Los Angeles Times*, 7/14/05.
63. Zerhouni, Elias A., Bethesda, MD, 1/24/07.
64. Klag, Michael J., Baltimore, MD, by telephone, 6/8/07.
65. Oldfield, Edward H., Charlottesville, VA, by telephone, 11/5/07.
66. Fischbach, Gerald D., New York, NY, by telephone, 3/8/07.
67. Shine, Kenneth I., Austin, TX, by telephone, 5/19/08.
68. Too strict at NIH. *The Washington Post*, 2/23/05.
69. Nathan, David G., Boston, MA, by telephone, 6/8/07.
70. Hall, Zach W., Wilson, WY, by telephone, 12/4/07.
71. Weiss, Rick. Pressure is building on NIH to reconsider conflict rules. *The Washington Post*, 4/17/05.

72. Notkins, A.L. Erosion of freedom of inquiry. *The NIH Catalyst* 2006;Jul.-Aug.:14–15.
73. Connolly, Ceci. Director of NIH agrees to loosen ethics rules. *The Washington Post*, 8/26/05.
74. Beamish, Rita. NIH's new ethics rules lead some to ponder jumping ship. *The Washington Post*, 10/30/06.
75. Ayres, Elaine J., Bethesda, MD, by telephone, 5/18/07.
76. Fleisher, Thomas A., Bethesda, MD, by e-mail, 4/14/08.
77. James, Stephen P., Bethesda, MD, by telephone, 8/1/07.
78. Common-sense ethics at NIH. *The Washington Post*, 8/28/05.
79. Relman, Arnold S., Boston, MA, by telephone, 2/19/07.
80. Corn, Milton, Bethesda, MD, by telephone, 10/30/07.
81. Lenardo, Michael J., Bethesda, MD, by telephone, 5/16/07.
82. Finkel, Toren, Bethesda, MD, by telephone, 10/2/07.
83. Paul, William E., Bethesda, MD, by telephone, 11/11/07.
84. Angell, M. *The truth about the drug companies. How they deceive us and what to do about it.* New York: Random House, 2005.
85. Angell, Marcia, Cambridge, MA, by telephone, 3/9/07.
86. Harris, Gardner, Carey, Benedict. Researchers fail to reveal full drug pay. *The New York Times*, 6/8/08.
87. Harris, Gardner. Leading psychiatrist failed to report drug income. *The New York Times*, 10/4/08.
88. Congressional Record—Senate. Payment to Physicians. Available at http://frwebgate.access.gpo.gov/cgi-bin/multidb.cgi?WAISdbName=2008_record+Congressional+Record%2C+Volume+154+%282008%29andWAISqueryRule=%28%24WAISqueryString%29andWAISqueryString=%22Grassley%22+and+%22Zerhouni%22andWAIStemplate=multidb_results.htmlandSubmit.=SubmitandWrapperTemplate=crecord_wrapper.htmlandWAISmaxHits=40. 6/4/2008, page S5029.
89. Kaiser, J. Senate inquiry on research conflicts shifts to grantees. *Science* 2008;320:1708.
90. Kaiser, J. More political heat on NIH. *Science* 2008;320:1407b.
91. Ruiz Bravo, Norka, Bethesda, MD, by telephone, 6/12/08.
92. Fredrickson, D.S. *The recombinant DNA controversy: A memoir. Science, politics, and the public interest 1974–1981.* Washington, DC: ASM Press, 2001.
93. Harden, Victoria A., Bethesda, MD, by telephone, 2/27/07.
94. Thomson, J.A., Itskovitz-Eldor, J., Shapiro, S.S., et al. Embryonic stem cell lines derived from human blastocysts. *Science* 1998;282:1145–1147.
95. Lefkowitz, Jay, P. Stem cells and the president. An inside account. *Commentary* 2008;Jan.:19–24. Available at http://www.commentarymagazine.com/viewpdf.cfm?article_id=11024.
96. Varmus, Harold E., New York, NY, by e-mail, 2/4/08.

97. Battey, Jr., James F., Bethesda, MD, by telephone, 5/14/07.
98. Marshall, E. View from the top of a biomedical empire. Interview by Eliot Marshall. *Science* 1999;285:1654–1656.
99. Chang, H.Y., Cotsarelis, G. Turning skin into embryonic stem cells. *Nature Medicine* 2007;13:783–784.
100. NIH Stem Cell Task Force, 2007. Available at http://stemcells.nih.gov/policy/taskforce/.
101. Alonso-Zaldivar, Ricardo, Kaplan, Karen. Loosening of stem cell limits backed. *Los Angeles Times*, 3/20/07.
102. Stolberg, Sheryl Gay. Obama lifts Bush's strict limits on stem cell research. *The New York Times*, 3/10/09.

Chapter 9: Conclusions

1. Gottesman, Michael M., Bethesda, MD, by telephone, 11/17/08.
2. Harris, Gardner. Specter, a fulcrum of the stimulus bill, pulls off a coup for health money. *The New York Times*, 2/14/09.

INDEX

Note: The letter "*n*" refers to notes.

Acupuncture, use of, 140
Administrative structure, 211–12
Adventist Hospital, 58
Advisory Board for Clinical
 Research (ABCR), 124–5
Advisory councils and committees,
 46–48
Advocates, 155–57
AIDS research, 61–64
 Vaccine Research Center, 64–67
Alberts, Bruce, 169, 196
Alexander, Duane, 82
Alter, Harvey, 127–28, 174
Alving, Barbara, 133
American Association of
 Universities, 158
Anfinsen, Christian, 24
Angell, Marcia, 204
Arai, Andrew, 60
Asian ginseng, use of, 140
Associate Training Program, 18
Association of American Medical
 Colleges (AAMC), 158
Augustine, Norman, 196
Autism, 73–74
Axelrod, Julius, 24
AZT (azidothymidine), 62

Baltimore, David, 28–29, 91, 93, 100,
 110–11
Barros, Colleen, 148, 152

Barton, Joe, 196–97
Battey, James, 96–97, 195, 208
Baum, Stanley, 102–3
Benz, Edward, 26–27, 39, 124
Berg, Jeremy, 83, 84–85
Big Pharma, 192
Bishop, J. Michael, 39
Black cohosh, use of, 141
Blue-ribbon panel, 195–97
Blumberg, Baruch, 127
Boards of Scientific Counselors
 (BSC), 10
Branch Chief, 8
Braunwald, Eugene, 4, 21–22, 21*n*,
 37, 56–57, 60
Bravo, Norka Ruiz, 112, 148, 149,
 183, 205
Briggs, Josephine, 139–40, 189
Broder, Samuel, 23, 62
Brown, Michael, 22, 22*n*, 36, 109,
 146, 180*n*
Brownback, Sam, 175–76
Brugge, Joan Siefert, 147, 158
Budget, xiv, 26–27, 142, 143, 180.
 See also Finances
 for Office of the Director (OD),
 163
Bureaucracy, 9–10
Bush, George W., 32, 52, 64, 148,
 162, 163, 168, 176, 206
Bush, Vannevar, 35

Butler, Robert, 93
Bypass budget, 49

Califf, Robert, 35
Cambridge/Oxford program, 13
Caplan, Arthur, 178, 193
Cardiac surgery program, 58, 59
Cardiovascular Sciences (CVS), 106
Careers, at NIH, 4–7
Cassell, Gail, 26
CDC. *See* Communicable Disease Center (CDC)
Cell, 205*n*. *See also* Stem cells
Center for Information Technology (CIT), 134–35
Center for Scientific Review (CSR), 105
    computerized applications, 116–17
    division and program directors, 112–14
    peer-review system, 109–12, 114–16
    study sections, 106–7
Centers, 105. *See also specific centers*
Centers for Disease Control and Prevention, 166. *See also* Communicable Disease Center (CDC)
Certificate of need (CON), 58
Charney, Dennis, 28
Clinical and Translational Science Awards (CTSA), 182–86
Clinical Center (CC), 18–19, 117–18
    Department of Laboratory Medicine, 126
    education, 125–26
    ethics, 132
    fibrous dysplasia of bone, investigating, 130–32
    management, 121–23
    patients, 118–21
    surgery, 128–29

Clinical Director, responsibilities of, 43
Clinical electives program, 13
Clinical research training program (CRTP), 13
Clinton, Bill, 32*n*, 100, 162, 164*n*, 168, 176, 207, 213
Cloister Program. *See* HHMI/NIH Research Scholars Program
Cloning, human, 207
Cloning of animals, 207*n*
Cochlear implants, 97–98
Collins, Francis, 27, 42, 99, 100, 101, 163, 167, 172, 175, 186, 191*n*
Collins, Michael, 130, 131, 132
Commissioned Corps, 18–19
Common Fund, 163, 180–81
    for the NIH director, 178–79
Communicable Disease Center (CDC), 19. *See also* Centers for Disease Control and Prevention
Conflict-of-interest controversy, 203
Conflicts of interest, in extramural program, 204
    director's response, 208–9
    NIH stem cell task force, 208
    Obama policy, 210
    politics and stem cell research, 206–8
    stem cells, 205–6
Conflicts of interest, in intramural program, 192–204
    blue-ribbon panel, 195–97
    degree of conflict of interest, 195
    new rules, 197–201
    rules, 192–93
    Trey Sunderland case, 193–95
    trouble, NIH in, 203–4
    Zerhouni's revision of rules, 201–3
Congressional friends, 154–55

Controversies, 192
  extramural program, conflicts of interest in, 204–10
  intramural program, conflicts of interest in, 192–204
Cooperative Research and Development Agreement (CRADA), 151
Corn, Milton, 174, 204
"Council of Councils", 179
Council of Public Representatives (COPR), 48
Croyle, Robert, 50, 52
Culliton, Barbara, 143, 156, 189

Danforth, William, 169n
Department of Transfusion Medicine, 127–28
Deputy Director, responsibilities of, 43
Deputy ethics counselor (DEC), 132
DeVita, Vincent, 36, 38, 110, 129n, 200–1
DeWitt Stetten, Jr., Museum of Medical Research, 17
Diabetes, NIDDK study in, 67–68
Director of Clinical Programs, responsibilities of, 43
Director of Extramural Activities (DEA), responsibilities of, 43–44
Director of Extramural Programs, responsibilities of, 44
Director of Intramural Activities, responsibilities of, 43
Directors, 161
  Healy, Bernadine, 163–69
  Office of the Director (OD), 163
  responsibilities, 43
  Varmus, Harold, 169–77
  Zerhouni, Elias, 177–91
Division of Extramural Activities Support (DEAS), 149

Division of Program Coordination, Planning, and Strategic Initiatives (DPCPSI), 179
DNA, 98
Doctor's Draft Law, 18n
Doubling, of budget, 143–44

Echinacea, use of, 141
Electronic Research Administration (eRA), 116
Emanuel, Ezekiel, 132, 199n, 201
Embryo, 206n
Embryonic stem cells, 206, 208
Eunice Kennedy Shriver National Institute of Child Health and Human Development (NICHD), 81–82
Extramural research program, 3. *See also* Intramural research program
  blaming, stopping, 39–40
  clinical research versus basic research, 35–39
  conflicts of interest in. *See* Conflicts of interest, in extramural program
  versus intramural research program, 212–13
  rescuing the NIH, 31–35

FASEB (Federation of American Societies for Experimental Biology), 155
Fauci, Anthony, 5, 6, 24, 26, 61, 162, 163, 164, 172, 176
Federally funded research and development center (FFRDC), 50, 55
Federation of American Societies for Experimental Biology, 155
Fellows, 13
Fibrous dysplasia of bone, investigating, 130–32
15-year project, 165

Finances, 142, 213–14.
    See also Budget
  consolidation and outsourcing,
    148–50
  doubling, 143–45
  salaries, 152–53
  straitened NIH, 145–48
  technology transfer, 150–51
Finkel, Toren, 57
Fischbach, Gerald, 43, 75–76, 156,
    170, 176–76, 201, 203, 207
Fisher, Suzanne, 108
FitzGerald, Garret, 185
Fogarty, John, 135–36
Foundation for Advanced
    Education in the Sciences
    (FAES), 16–17
Fraser-Liggett, Claire, 42n
Fredrickson, Donald, 24, 188n
Funds, for NIH, 154

Gage, Fred, 209
Gajdusek, Carleton, 24
Gallin, John, 14, 24, 58, 117, 119, 120,
    121, 122, 123, 124, 126
Gallo, Robert, 9, 23, 29, 62, 65
GenBank®, 80
General Clinical Research Centers
    (GCRC), 182, 184
Gibbons, Don, 157
Gingrich, Newt, 143, 155
Ginkgo, use of, 141
Glass, Roger, 136
Glucosamine plus chondroitin
    sulfate, use of, 140
Goal-directed grants, 110
Goldstein, Joseph, 22n, 28
Goldman, Lee, 25n
Gorden, Phillip, 24, 177n
Gottesman, Michael, 10, 11, 14, 15, 16,
    24, 29, 49, 99–100, 167, 204
Graduate degrees, 14–16
Graduate partnership program
    (GPP), 13, 15

Graduate partnership program with
    Cambridge and Oxford
    Universities, 13
Grady, Patricia, 95, 172
Graeff, Alan, 134–35
Gragnolati, Brian, 59
Grants.gov website, 116
Grassley, Charles, 204
Greenberg, Daniel, 46
Greenberger, Phyllis, 44n
Greenwood, James, 196–97
GRID (gay related immune
    deficiency), 61, 62
Group of Concerned Universities
    and Research Institutions,
    157–60

Hall, Zach, 74–75
Harden, Victoria, 161, 167,
    167n, 174
Harkin, Tom, 137, 143, 201, 209
Harvard University, 157
Hastings, Clare, 120
Healy, Bernadine, 24, 93, 99, 162,
    163–65, 188n
  criticism, 168
  opinions, 167–68
  post-NIH activities, 169
  projects, 165–67
Henderson, David, 124
HHMI/NIH Research Scholars
    Program, 12
History, of NIH, xv–xvi
HIV/AIDS, research in,
    61–65
Hodes, Richard, 24, 93, 168,
    172, 181
Holden, Kenneth, 167
Horvath, Keith, 59
Human-based studies, performance
    of, 37
Human cloning, 207
Human Genome Project, 98, 99
Human resources (HR), 148

Hygienic Laboratory, xv
Hyman, Steven, 29, 71–72, 156, 162

Index Medicus®, 79
Insel, Thomas, 72–73
Institute of Medicine's (IOM) 2003 report, 162
Institutes, 41. *See also specific institutes*
  advisory councils and committees, 46–48
  directors and executives, 43–44
  leadership, 41–43
  number of, 44–46
Institutional review boards (IRBs), 37, 120n
Integrated Review Group (IRG), 106
Intergovernmental Personnel Act (IPA), 60
Internet, in National Library of Medicine, 79–80
Intramural research program (IRP), 3–4, 163. *See also* Extramural research program
  careers, 4–7
  Commissioned Corps, 18–19
  conflicts of interest in. *See* Conflicts of interest, in intramural program
  current status, 24–30
  versus extramural research program, 212–13
  Foundation for Advanced Education in the Sciences (FAES), 16–17
  graduate degrees, 14–16
  Office of NIH History, 17
  problems about working in, 8–10
  reviews, of staff investigators' works, 10–11
  staff positions, 7–8
  training, 12–14
  Yellow Berets, 20–24

Investigators, 7–8
  rank. *See* Staff positions
  tenure-track, 7
In vitro fertilization, 206

Jacobs, Joseph, 138–39
James, Stephen, 108, 112, 202n
John E. Fogarty International Center (FIC), 135–36

Kandel, Eric, 15
Katz, Stephen, 94, 114
Kelley, William, 170
Killen, John "Jack," 139
Kinyoun, Joseph J., xv
Kirschner, Marc, 15, 161
Kirschstein, Ruth, 83, 84, 139, 148, 169n, 176, 177, 177n
Klausner, Richard, 49, 51–52, 110, 195
Klein, Harvey, 23, 120, 127
Klein, Melissa, 20
Koop, C. Everett "Chick," 20n
Korn, David, 3, 39, 146
Kornberg, Arthur, 9, 180n
Krensky, Alan, 186
Kupfer, Carl, 89

Lab Chief, 8
Lander, Eric, 100
Landis, Story, 77–78, 181
Lauer, Michael, 6
Lee, Philip, 143, 169n, 174
Lefkowitz, Jay, 206n
Lenfant, Claude, 56n
Leshner, Alan, 91–92, 169n, 172, 174
Levine, Arthur, 81, 146
Levitt, Michael, 201
Lister Hill National Center for Biomedical Communications (LHNCBC), 78
Loop, Floyd, 164n
Lott, Trent, 103
Lunacy, 166

Mack, Connie, 143
Marburger, John, 159
Marine Hospital Service (MHS), xv
Marks, Andrew, 31–32, 35
Martensen, Robert, 17
Martin, Joseph, 158
Maryland Health Care
  Commission, 58
Masur, Henry, 22
Mayo Clinical Laboratories, 126
McKhann, Guy, 8–9, 23, 41–42,
  170, 188
McNeil, Jr., Robert L., 185
Medical Scientist Training Program
  (MSTP), 85–86
MEDLINE databases, 79
MedlinePlus, 80
MedStar Health, 58
Mega-silos, 41
Membership, in councils, 47
Memorial Sloan-Kettering Cancer
  Center (MSKCC), 16n
MHS. See Marine Hospital Service
  (MHS)
Mikulski, Barbara, 165
Miller, Edward, 178, 180n
Molecular libraries, 186
Moonlighting, 202n
Morrow, Glenn, 57
Multiple-PI model, 183

Nabel, Elizabeth (Betsy), 6, 56,
  171n
Nabel, Gary, 64
Nathan, David, 184, 201
National Alliance on Mental Illness
  (NAMI), 156
National Cancer Advisory
  Board, 49
National Cancer Institute (NCI),
  xv–xvi, 11, 48, 138, 150, 211
  administration, 49–50
  Klausner, Richard, 51–52
  NCI–Frederick, 50–51

Niederhuber, John, 54–55
von Eschenbach, Andrew,
  52–54
National Cancer Program, 49
National Center for Biotechnology
  Information (NCBI), 78
National Center for
  Complementary and
  Alternative Medicine
  (NCCAM), 137–41
National Center for Research
  Resources (NCRR), 133–34,
  184
National Center on Minority
  Health and Health Disparities
  (NCMHD), 136–37
National Children's Study, 82
National Diabetes Education
  program (NDEP), 67
National Eye Institute (NEI), 89
National Health Council, 157
National Heart, Lung, and Blood
  Institute (NHLBI), 55–56,
  150, 211
  clinical programs, 56–58
  Suburban Hospital, 58–60
National Human Genome
  Research Institute (NHGRI),
  97–101
National Institute of Alcohol
  Abuse and Alcoholism
  (NIAAA), 69
National Institute of Allergy and
  Infectious Diseases (NIAID),
  61–65, 150
National Institute of Arthritis,
  Musculoskeletal, and Skin
  Diseases, 211
National Institute of Arthritis and
  Metabolic Diseases (NIAMD),
  66
National Institute of Arthritis and
  Musculoskeletal and Skin
  Diseases (NIAMS), 67, 93–94

National Institute of Biomedical
  Imaging and Bioengineering
  (NIBIB), 102-4
National Institute of Dental and
  Craniofacial Research
  (NIDCR), 65-66
National Institute of Diabetes and
  Digestive and Kidney Diseases
  (NIDDK), 66-67, 150, 211, 212
  diabetes, 67-68
  Polycystic Kidney Disease
  (PKD), 68-69
National Institute of Environmental
  Health Sciences (NIEHS),
  86-88
National Institute of General
  Medical Sciences (NIGMS),
  83-84
  function and purpose, 84-85
  grants, funding, 85
  Medical Scientist Training
  Program, 85-86
National Institute of Mental Health
  (NIMH), 69-73
National Institute of Neurological
  and Communicative Disorders
  and Stroke (NINCDS), 74
National Institute of Neurological
  Disorders and Stroke
  (NINDS), 74-78
National Institute of Nursing
  Research (NINR), 65-66,
  94-95
National Institute on Aging (NIA),
  92-93
National Institute on Alcohol Abuse
  and Alcoholism (NIAAA),
  90-91
National Institute on Deafness and
  Other Communication
  Disorders (NIDCD),
  74, 95-97
National Institute on Drug Abuse
  (NIDA), 69, 91-92

National Institutes of Health and the
  Food and Drug
  Administration, 151
National Institutes of Health
  Reform Act of 2006, 178-79
National Interagency Biodefense
  Campus (NIBC), 50n
National Library of Medicine
  (NLM), 78-79
  internet, 79-80
  public-access policy, 80
NCI. *See* National Cancer Institute
  (NCI)
NCI-Frederick, 50-51
New Innovator Awards, 186
Niederhuber, John, 54-56
Nirenberg, Marshall, 24, 56
Nixon, Richard, 156
Notkins, Abner, 20-21, 21n,
  199, 201
Nurse, Paul, 54

Obama, Barack, 210, 213
Office of Management and Budget
  (OMB), 148
Office of NIH History, 17
Office of Portfolio Analysis and
  Strategic Initiatives (OPASI),
  186-87
Office of Technology Transfer
  (OTT), 150
Office of the Director (OD), 163
Oldfield, Edward, 28
Open society, of NIH, 193
Oxford/Cambridge program, 13

Pardes, Herbert, 4, 70
Passamani, Eugene, 59
Patenting human genes, 166
Paul, William, 63, 64
Pay-lines, 46
Penn program, 185
Pettigrew, Roderic, 104
Physician-investigators, 34

Pioneer Awards, 186
Pizzo, Philip, 5, 25, 62
Polycystic Kidney Disease (PKD), 68–69
Porter, John Edward, 71, 74, 143, 144, 146
Porter Neuroscience Research Center, 74
Post- baccalaureate ("post-bac") intramural training award, 12
Postdoctoral program, 13
Potts, John, 20, 20n
Principal investigators (PI), 7
Private companies, NIH collaboration with, 150
Public-access policy of the NIH, 80
Public Health Service (PHS), 18n
Public recognition, of NIH, 166
PubMed, 79
PubMed Central (PMC), 79, 80

R01, 11n
Reagan, Ronald, 70, 205n
Recombinant DNA, 205n
Relman, Arnold "Bud", 5–6, 203
Research and Clinical Fellows, 7
Reviews, of staff investigators' works, 10–11
Reynolds, Craig, 51
Roadmap for Medical Research, 179–82
Rodbell, Martin, 24
Rodgers, Griffin, 37
Rodin, Judith, 169n
Rogers, Paul, 155, 189
Rohrbaugh, Marc, 151
Rosenberg, Steven, 120, 128–29
Roth, Carl, 203
Ruiz Bravo, Norka, 33, 34

Salaries, 152–53
Scarpa, Antonio (Toni), 107, 108, 109

Schechter, Alan, 39, 171, 173, 180, 182n
Schimpff, Stephen, 121
Schroeder, Pat, 165
Schwartz, David, 87–88
Science Education Partnership Awards (SEPA), 134
Scientific Interest Groups, 173
Scientific Management Review Board, 179
Scientific review officer (SRO), 106
Senior Investigators, 8
Shalala, Donna, 92, 121–22, 168, 169, 172
Shamoo, Adil, 111
Shannon, James, 44n, 161, 165–66
Shine, Kenneth, 36, 143, 162, 172, 200, 204
Sieving, Paul, 28, 180
Singer, Daniel, 185
Sloan-Kettering, 175
Spaeth, Jennifer, 48
Specter, Arlen, 143, 155
Spinal manipulation for lower back pain, use of, 140
Staff positions, at NIH, 7–8
  Research and Clinical Fellows, 7
  Investigators, 7–8
  Senior Investigators, 8
  unofficial designations, 8
  Staff Scientists and Staff Clinicians, 7
Steering committee, members of, 187
Stem cells, 205
  research in, 206–8
  task force, 208
St. John's wort, use of, 140
Straitened NIH, 145–48
Straus, Stephen, 139
Substance Abuse and Mental Health Services Administration (SAMHSA), 69–70

Suburban Hospital, 58–60
Sullivan, Louis, 164, 140
Summer internship program, 12
Summer research fellowship
   program, 12
Sunderland, Trey, 193–95

Tabak, Lawrence, 65
Technology transfer, 150–51
Tenure, of staff, 7
Thier, Samuel, 40
Thompson, Tommy, 148
Tilghman, Shirley, 15, 23
Title 5, 152
Title 42, 152
Trey Sunderland case, 193–95

Vaccine Research Center, 64–67
Vagelos, Roy, 38, 190
Varmus, Harold, 14–15, 23, 24, 26,
   38, 39, 42, 44–45, 48, 52, 58, 63,
   64, 67, 71, 74–75, 76, 91, 93,
   95, 96, 103, 114, 121, 123, 142,
   143, 156, 159, 162, 165, 171n,
   164, 175n, 179, 180, 192, 193,
   199, 200, 207, 209, 212
   departure, 174–75
   praise and criticism, 173–74
   as transplanted Westerner, 171–73

Venter, J. Craig, 166
Volkow, Nora, 92
Von Eschenbach, Andrew, 52–54

Washington Hospital Center, 58
Watson, James, 98, 99, 166
Weinberg, Myrl, 157
Weissmann, Gerald, 33, 181
Wildenthal, Kern, 146
Wiltrout, Robert, 204
Woolley, Mary, 156–57
Wyatt, Richard, 10
Wyngaarden, James, 24, 99, 188n

Yamamoto, Keith, 190
Year-off training program, 12
Yellow Berets, 20–24

Zerhouni, Elias, 5n, 24n, 32, 35, 37–38,
   57–58, 83, 110, 147, 151, 153, 156,
   162, 175, 177–91
   background, 188
   leaving NIH, 190–91
   governance, 187–88
   National Institutes of Health
      Reform Act of 2006, 178–79
   opinions, 189–90
   Roadmap for Medical Research,
      179–82

www.ingramcontent.com/pod-product-compliance
Ingram Content Group UK Ltd.
Pitfield, Milton Keynes, MK11 3LW, UK
UKHW022153230426
12049UKWH00003BA/68